T0212378

SpringerBriefs in Optimization

Series Editors

Sergiy Butenko, Department of Industrial and Systems Engineering, Texas A&M University, College Station, TX, USA

Mirjam Dür, Department of Mathematics, University of Trier, Trier, Germany

Panos M. Pardalos, ISE Department, University of Florida, Gainesville, FL, USA

János D. Pintér, Lehigh University, Bethlehem, PA, USA

Stephen M. Robinson, University of Wisconsin-Madison, Madison, WI, USA

Tamás Terlaky, Lehigh University, Bethlehem, PA, USA

My T. Thai ⓘ, CISE Department, University of Florida, Gainesville, FL, USA

SpringerBriefs present concise summaries of cutting-edge research and practical applications across a wide spectrum of fields. Featuring compact volumes of 50 to 125 pages, the series covers a range of content from professional to academic. Briefs are characterized by fast, global electronic dissemination, standard publishing contracts, standardized manuscript preparation and formatting guidelines, and expedited production schedules.

Typical topics might include

- A timely report of state-of-the art techniques
- A bridge between new research results, as published in journal articles, and a contextual literature review
- A snapshot of a hot or emerging topic
- An in-depth case study
- A presentation of core concepts that students must understand in order to make independent contributions

SpringerBriefs in Optimization showcase algorithmic and theoretical techniques, case studies, and applications within the broad-based field of optimization. Manuscripts related to the ever-growing applications of optimization in applied mathematics, engineering, medicine, economics, and other applied sciences are encouraged.

More information about this series at http://www.springer.com/series/8918

Alexander J. Zaslavski

The Projected Subgradient Algorithm in Convex Optimization

 Springer

Alexander J. Zaslavski
Department of Mathematics
Technion – Israel Institute of Technology
Haifa, Israel

ISSN 2190-8354 ISSN 2191-575X (electronic)
SpringerBriefs in Optimization
ISBN 978-3-030-60299-4 ISBN 978-3-030-60300-7 (eBook)
https://doi.org/10.1007/978-3-030-60300-7

Mathematics Subject Classification: 49M37, 65K05, 90C25, 90C26, 90C30

This Springer imprint is published by the registered company Springer Nature Switzerland AG
The registered company address is: Gewerbestrasse 11, 6330 Cham, Switzerland

Contents

1	**Introduction**	1
	1.1 Subgradient Projection Method	1
2	**Nonsmooth Convex Optimization**	5
	2.1 Preliminaries	5
	2.2 Approximate Solutions	7
	2.3 Convergence to the Solution Set	18
	2.4 Superiorization	22
	2.5 Auxiliary Results for Theorems 2.1–2.10	24
	2.6 Proof of Theorem 2.1	35
	2.7 Proof of Theorem 2.2	36
	2.8 Proof of Theorem 2.3	39
	2.9 Proof of Theorem 2.5	41
	2.10 Proof of Theorem 2.8	43
	2.11 Proof of Theorem 2.10	45
	2.12 An Auxiliary Result for Theorems 2.11–2.15	48
	2.13 Proof of Theorem 2.11	49
	2.14 Proof of Theorem 2.12	54
	2.15 Proof of Theorem 2.13	59
	2.16 Proof of Theorem 2.14	65
	2.17 Proof of Theorem 2.15	72
	2.18 Proof of Theorem 2.16	78
	2.19 Proof of Theorem 2.17	80
	2.20 Proof of Theorem 2.18	82
3	**Extensions**	85
	3.1 Optimization Problems on Bounded Sets	85
	3.2 An Auxiliary Result for Theorem 3.2	89
	3.3 An Auxiliary Result for Theorem 3.3	91
	3.4 Proof of Theorem 3.2	94
	3.5 Proof of Theorem 3.3	95
	3.6 Optimization on Unbounded Sets	96

3.7	Auxiliary Results	100
3.8	Proof of *Theorem 3.6	106
3.9	Proof of Theorem 3.7	109
4	**Zero-Sum Games with Two Players**	**113**
4.1	Preliminaries and an Auxiliary Result	113
4.2	Zero-Sum Games on Bounded Sets	117
5	**Quasiconvex Optimization**	**129**
5.1	Preliminaries	129
5.2	The Main Lemma	133
5.3	Optimization on Bounded Sets	134
5.4	Optimization on Unbounded Sets	137
References		**143**

Chapter 1
Introduction

In this book we study the behavior of subgradient algorithms for constrained minimization problems in a Hilbert space. Our goal is to obtain a good approximate solution of the problem in the presence of computational errors. It is known that the algorithm generates a good approximate solution, if the sequence of computational errors is bounded from above by a small constant. In our study, presented in this book, we take into consideration the fact that for every algorithm, its iteration consists of several steps and that computational errors for different steps are different, in general. In this section we discuss several algorithms which are studied in the book.

1.1 Subgradient Projection Method

In this book we study the subgradient projection algorithm for minimization of convex and nonsmooth functions and for computing the saddle points of convex-concave functions, under the presence of computational errors. It should be mentioned that the subgradient projection algorithm is one of the most important tools in the optimization theory, nonlinear analysis, and their applications. See, for example, [1–3, 7, 10–12, 16, 17, 26, 28–30, 33–35, 37, 43–47, 50–56, 58–60, 64–68, 71–75, 77, 78]. The problem is described by an objective function and a set of feasible points. For this algorithm, each iteration consists of two steps. The first step is a calculation of a subgradient of the objective function, while in the second one, we calculate a projection on the feasible set. In each of these two steps, there is a computational error. In general, these two computational errors are different. In our recent research [77], we show that our algorithm generates a good approximate solution, if all the computational errors are bounded from above by a small positive constant. Moreover, if we know computational errors for the two steps of our algorithm, we find out what an approximate solution can be obtained

A. J. Zaslavski, *The Projected Subgradient Algorithm in Convex Optimization*,
SpringerBriefs in Optimization, https://doi.org/10.1007/978-3-030-60300-7_1

and how many iterates one needs for this. In this book we generalize all these results for an extension of the projected subgradient method, when instead of the projection on the feasible set it is used a quasi-nonexpansive retraction on this set.

We study the subgradient algorithm for constrained minimization problems in Hilbert spaces equipped with an inner product denoted by $\langle \cdot, \cdot \rangle$ which induces a complete norm $\| \cdot \|$ and use the following notation.

For every $z \in R^1$ denote by $\lfloor z \rfloor$ the largest integer which does not exceed z:

$$\lfloor z \rfloor = \max\{i \in R^1 : i \text{ is an integer and } i \leq z\}.$$

For every nonempty set D, every function $f : D \to R^1$ and every nonempty set $C \subset D$, we set

$$\inf(f, C) = \inf\{f(x) : x \in C\}.$$

Let X be a Hilbert space equipped with an inner product denoted by $\langle \cdot, \cdot \rangle$ which induces a complete norm $\| \cdot \|$. For each $x \in X$ and each $r > 0$, set

$$B_X(x, r) = \{y \in X : \|x - y\| \leq r\},$$

and for each $x \in X$ and each nonempty, set $E \subset X$ set

$$d(x, E) = \inf\{\|x - y\| : y \in E\}.$$

For each nonempty open convex, set $U \subset X$ and each convex function $f : U \to R^1$; for every $x \in U$, set

$$\partial f(x) = \{l \in X : f(y) - f(x) \geq \langle l, y - x \rangle \text{ for all } y \in U\}$$

which is called the subdifferential of the function f at the point x [48, 49, 61].

Let C be a nonempty closed convex subset of X and let $f : X \to R^1$ be a convex function.

Suppose that there exist $L > 0$, $M_0 > 0$ such that

$$C \subset B_X(0, M_0),$$

$$|f(x) - f(y)| \leq L\|x - y\| \text{ for all } x, y \in B_X(0, M_0 + 2).$$

It is not difficult to see that for each $x \in B_X(0, M_0 + 1)$,

$$\emptyset \neq \partial f(x) \subset B_X(0, L).$$

It is well-known that for every nonempty closed convex, set $D \subset X$ and every $x \in X$, there is a unique point $P_D(x) \in D$ satisfying

$$\|x - P_D(x)\| = \inf\{\|x - y\| : y \in D\}.$$

We consider the minimization problem

$$f(z) \rightarrow \min, \ z \in C.$$

Suppose that $\{\alpha_k\}_{k=0}^{\infty} \subset (0, \infty)$. Let us describe our algorithm.

Subgradient Projection Algorithm
Initialization: select an arbitrary $x_0 \in B_X(0, M_0 + 1)$.
Iterative step: given a current iteration vector $x_t \in U$, calculate $\xi_t \in \partial f(x_t)$ and the next iteration vector $x_{t+1} = P_C(x_t - \alpha_t \xi_t)$.

In [75] we study this algorithm under the presence of computational errors. Namely, in [75], we suppose that $\delta \in (0, 1]$ is a computational error produced by our computer system, and study the following algorithm.

Subgradient Projection Algorithm with Computational Errors
Initialization: select an arbitrary $x_0 \in B_X(0, M_0 + 1)$.
Iterative step: given a current iteration vector $x_t \in B_X(0, M_0 + 1)$, calculate $\xi_t \in \partial f(x_t) + B_X(0, \delta)$ and the next iteration vector $x_{t+1} \in U$ such that $\|x_{t+1} - P_C(x_t - a_t \xi_t)\| \leq \delta$.

In [77] we consider more complicated, but more realistic, version of this algorithm. Clearly, for the algorithm, each iteration consists of two steps. The first step is a calculation of a subgradient of the objective function f, while in the second one, we calculate a projection on the set C. In each of these two steps, there is a computational error produced by our computer system. In general, these two computational errors are different. This fact is taken into account in the following projection algorithm studied in Chapter 2 of [77].
Suppose that $\{\alpha_k\}_{k=0}^{\infty} \subset (0, \infty)$ and $\delta_f, \delta_C \in (0, 1]$.
Initialization: select an arbitrary $x_0 \in B_X(0, M_0 + 1)$.
Iterative step: given a current iteration vector $x_t \in B_X(0, M_0 + 1)$, calculate $\xi_t \in \partial f(x_t) + B_X(0, \delta_f)$ and the next iteration vector $x_{t+1} \in U$ such that $\|x_{t+1} - P_C(x_t - \alpha_t \xi_t)\| \leq \delta_C$.

Note that in practice for some problems, the set C is simple but the function f is complicated. In this case, δ_C is essentially smaller than δ_f. On the other hand, there are cases when f is simple but the set C is complicated, and therefore δ_f is much smaller than δ_C.

In our analysis of the behavior of the algorithm in [75, 77], properties of the projection operator P_C play an important role. In the present book, we obtain generalizations of the results obtained in [75, 77] for the subgradient methods in the case when the set C is not necessarily convex and the projection operator P_C is replaced by a mapping $P : X \rightarrow C$ which satisfies

$$Px = x \text{ for all } x \in C, \tag{1.1}$$

$$\|Px - z\| \leq \|x - z\| \text{ for all } z \in C \text{ and all } x \in X. \tag{1.2}$$

In other words, P is a quasi-nonexpansive retraction on C. Note that there are many mappings $P : X \to C$ satisfying (1.1) and (1.2). Indeed, in [57] we consider a space of mappings $P : X \to X$ satisfying (1.1) and (1.2), which is equipped with a natural complete metric, and show that for a generic (typical) mapping from the space, its powers converge to a mapping which also satisfies (1.1) and (1.2) and such that its image is C.

Note that the generalizations considered in this book have, besides their obvious mathematical interest, also a significant practical meaning. Usually, the projection operator $P_C : X \to C$ can be calculated when C is a simple set like a linear subspace, a half-space, or a simplex. In practice, C is an intersection of simple sets C_i, $i = 1, \ldots, q$, where q is a large natural number. The calculation of P_C is not possible in principle. Instead, it is possible to calculate the product $P_{C_q} \cdots P_{C_1}$ and its powers $(P_{C_q} \cdots P_{C_1})^m$, $m = 1, 2, \ldots$.

It is well-known [76] that under certain regularity conditions on C_i, $i = 1, \ldots, q$ the powers $(P_{C_q} \cdots P_{C_1})^m$ converge as $m \to \infty$ to a mapping $P : X \to C$ which satisfies (1.1) and (1.2). Thus in practice we cannot calculate the projection operator P_C but only a mapping $P : X \to C$ satisfying (1.1) and (1.2) [4, 5, 8, 9, 21, 24, 27, 31, 32, 36, 52, 62, 69] or, more exactly, its approximations. This shows that the results of this book are indeed important from the point of view of practice.

In Chapter 2, we study the subgradient projection algorithm presented above for convex minimization problems with objective functions defined on the whole Hilbert space. In Chapter 3, we generalize some results of Chapter 2 for the case of problems with objective functions defined on subsets of the Hilbert space. In Chapter 4, we study the subgradient projection algorithm for zero-sum games with two players. In Chapter 5, we study the projected subgradient method for quasiconvex optimization problems.

Chapter 2
Nonsmooth Convex Optimization

In this chapter, we study an extension of the projected subgradient method for minimization of convex and nonsmooth functions, under the presence of computational errors. The problem is described by an objective function and a set of feasible points. For this algorithm, each iteration consists of two steps. The first step is a calculation of a subgradient of the objective function, while in the second one, we calculate a projection on the feasible set. In each of these two steps, there is a computational error. In general, these two computational errors are different. In our recent research [77], we show that our algorithm generates a good approximate solution, if all the computational errors are bounded from above by a small positive constant. Moreover, if we know computational errors for the two steps of our algorithm, we find out what an approximate solution can be obtained and how many iterates one needs for this. In this chapter, we generalize all these results for an extension of the projected subgradient method, when instead of the projection on the feasible set it is used a quasi-nonexpansive retraction on this set.

2.1 Preliminaries

Let $(X, \langle \cdot, \cdot \rangle)$ be a Hilbert space with an inner product $\langle \cdot, \cdot \rangle$ which induces a complete norm $\| \cdot \|$. For each $x \in X$ and each nonempty set $A \subset X$, set

$$d(x, A) = \inf\{\|x - y\| : y \in A\}.$$

For each $x \in X$ and each $r > 0$, set

$$B_X(x, r) = \{y \in X : \|x - y\| \leq r\}.$$

A. J. Zaslavski, *The Projected Subgradient Algorithm in Convex Optimization*,
SpringerBriefs in Optimization, https://doi.org/10.1007/978-3-030-60300-7_2

Assume that $f : X \rightarrow R^1$ is a convex continuous function which is Lipschitz on all bounded subsets of X. For each point $x \in X$ and each positive number ϵ, let

$$\partial f(x) = \{l \in X : \ f(y) - f(x) \geq \langle l, y - x \rangle \text{ for all } y \in X\} \tag{2.1}$$

be the subdifferential of f at x [48, 49], and let

$$\partial_\epsilon f(x) = \{l \in X : \ f(y) - f(x) \geq \langle l, y - x \rangle - \epsilon \text{ for all } y \in X\} \tag{2.2}$$

be the ϵ-subdifferential of f at x.

Let C be a closed nonempty subset of the space X. Assume that

$$\lim_{\|x\| \to \infty} f(x) = \infty. \tag{2.3}$$

It means that for each $M_0 > 0$, there exists $M_1 > 0$ such that if a point $x \in X$ satisfies the inequality $\|x\| \geq M_1$, then $f(x) > M_0$.

In this chapter, we consider the optimization problem

$$f(x) \rightarrow \min, \ x \in C.$$

Define

$$\inf(f, C) = \inf\{f(z) : \ z \in C\}. \tag{2.4}$$

Since the function f is Lipschitz on all bounded subsets of the space X, it follows from (2.3) that $\inf(f, C)$ is finite. Set

$$C_{min} = \{x \in C : \ f(x) = \inf(f, C)\}. \tag{2.5}$$

It is well-known that if the set C is convex, then the set C_{min} is nonempty. Clearly, the set $C_{min} \neq \emptyset$ if the space X is finite-dimensional.

We assume that

$$C_{min} \neq \emptyset. \tag{2.6}$$

It is clear that C_{min} is a closed subset of X. Define

$$X_0 = \{x \in X : \ f(x) \leq \inf(f, C) + 4\}. \tag{2.7}$$

In view of (2.3), there exist a number $\bar{K} > 1$ such that

$$X_0 \subset B_X(0, \bar{K}). \tag{2.8}$$

Since the function f is Lipschitz on all bounded subsets of the space X, there exist a number $\bar{L} > 1$ such that

$$|f(z_1) - f(z_2)| \le \bar{L}\|z_1 - z_2\| \text{ for all } z_1, z_2 \in B_X(0, \bar{K} + 4). \tag{2.9}$$

Denote by $\mathcal{M}$ the set of all mappings $P : X \to X$ such that for all $x \in C$ and all $y \in X$,

$$\|x - Py\| \le \|x - y\|, \tag{2.10}$$

$$Px = x \text{ for all } x \in C. \tag{2.11}$$

For every $P \in \mathcal{M}$, set $P^0 x = x$, $x \in X$.

2.2 Approximate Solutions

In the following results, our goal is to obtain an approximate solution x which is close to the set C such that $f(x)$ is closed to $\inf(f, C)$. In the first theorem, the set C is bounded, the computational errors δ_f, δ_C are given, and the step-size α depends on δ_f, δ_C. It is proved in Section 2.6.

Theorem 2.1 *Assume that* $K_1 \ge \bar{K} + 1$, $L_1 \ge \bar{L}$, $\delta_f, \delta_C \in (0, 1]$,

$$C \subset B_X(0, K_1), \tag{2.12}$$

$$|f(z_1) - f(z_2)| \le L_1\|z_1 - z_2\| \text{ for all } z_1, z_2 \in B_X(0, K_1 + 2), \tag{2.13}$$

$$\delta_f(\bar{K} + K_1 + 2 + 5L_1 + 5\bar{L}_1) \le 1, \ \delta_C(\bar{K} + K_1 + 2 + 5L_1 + 5\bar{L}_1) \le 1, \tag{2.14}$$

$$\epsilon = \max\{48\delta_f(\bar{K} + K_1 + 2 + 5L_1 + 5\bar{L}), \ 40(L_1 + \bar{L})(\delta_C(\bar{K} + K_1 + 2 + 5L_1 + 5\bar{L}))^{1/2}\} \tag{2.15}$$

and that

$$n = \lfloor 800(1 + K_1 + \bar{K})^2(L_1 + \bar{L})^2\epsilon^{-2}\rfloor + 2. \tag{2.16}$$

Let $\{P_i\}_{i=0}^{n-1} \subset \mathcal{M}$ *satisfy*

$$P_i(X) = C, \ i = 0, \ldots, n - 1, \tag{2.17}$$

$$\alpha = 10^{-2}(L_1 + \bar{L})^{-2}\epsilon, \tag{2.18}$$

$$\{x_i\}_{i=0}^n \subset X, \ \{\xi_i\}_{i=1}^{n-1} \subset X,$$

$$\|x_0\| \leq K_1, \ \|x_1 - P_0 x_0\| \leq \delta_C \tag{2.19}$$

and that for $i = 1, \ldots, n - 1$,

$$B_X(\xi_i, \delta_f) \cap \partial_{\epsilon/4} f(x_i) \neq \emptyset, \tag{2.20}$$

$$\|x_{i+1} - P_i(x_i - \alpha \xi_i)\| \leq \delta_C. \tag{2.21}$$

Then there exists $j \in \{1, \ldots, n\}$ such that

$$f(x_j) \leq \inf(f, C) + \epsilon.$$

In the second theorem, the set C is not necessarily bounded, the computational errors δ_f, δ_C are given, and the step-size α depends on δ_f, δ_C. It is proved in Section 2.7.

Theorem 2.2 *Assume that* $K_1 \geq \bar{K} + 1$, $L_1 \geq \bar{L}$, $\delta_f, \delta_C \in (0, 1]$,

$$|f(z_1) - f(z_2)| \leq L_1 \|z_1 - z_2\| \text{ for all } z_1, z_2 \in B_X(0, 3K_1 + 2), \tag{2.22}$$

$$\delta_f(\bar{K} + 3K_1 + 2 + 5L_1 + 5\bar{L}) \leq 1, \tag{2.23}$$

$$\delta_C(\bar{K} + 3K_1 + 2 + 5L_1 + 5\bar{L}) \leq (10(\bar{L} + L_1))^{-2}, \tag{2.24}$$

$$\begin{aligned}
\epsilon = \max\{&4\delta_f(\bar{K} + 3K_1 + 2 + 5L_1 + 5\bar{L}), \\
&40(L_1 + \bar{L})(\delta_C(\bar{K} + 3K_1 + 2 + 5L_1 + 5\bar{L}))^{1/2}\}
\end{aligned} \tag{2.25}$$

and that

$$n = \lfloor 800(1 + K_1 + \bar{K})^2 (L_1 + \bar{L})^2 \epsilon^{-2} \rfloor + 2. \tag{2.26}$$

Let $\{P_i\}_{i=0}^{n-1} \subset \mathcal{M}$ *satisfy*

$$P_i(X) = C, \ i = 0, \ldots, n - 1, \tag{2.27}$$

$$\alpha = 10^{-2}(L_1 + \bar{L})^{-2}\epsilon, \tag{2.28}$$

$\{x_i\}_{i=0}^{n} \subset X$, $\{\xi_i\}_{i=1}^{n-1} \subset X$,

$$\|x_0\| \leq K_1, \tag{2.29}$$

$$\|x_1 - P_0 x_0\| \leq \delta_C \tag{2.30}$$

and that for all $i = 1, \ldots, n - 1$,

$$B_X(\xi_i, \delta_f) \cap \partial_{\epsilon/4} f(x_i) \neq \emptyset, \tag{2.31}$$

$$\|x_{i+1} - P_i(x_i - \alpha\xi_i)\| \leq \delta_C. \tag{2.32}$$

Then there exists $j \in \{1, \ldots, n\}$ such that

$$f(x_j) \leq \inf(f, C) + \epsilon.$$

In the third theorem, the set C is bounded, the computational errors δ_f, δ_C and ϵ are given, and the step-size $\alpha_i, i = 1, \ldots, n - 1$ are given too. It is proved in Section 2.8.

Theorem 2.3 *Assume that $K_1 \geq \bar{K} + 1$, $L_1 \geq \bar{L}$, $\delta_f, \delta_C \in (0, 1]$,*

$$C \subset B_X(0, K_1), \tag{2.33}$$

$$|f(z_1) - f(z_2)| \leq L_1\|z_1 - z_2\| \text{ for all } z_1, z_2 \in B_X(0, K_1 + 2), \tag{2.34}$$

$$0 < \epsilon \leq 16(\bar{L} + L_1 + 1) \tag{2.35}$$

and that

$$\delta_f(\bar{K} + K_1 + 2 + 5L_1 + 5\bar{L}_1) \leq \epsilon/8. \tag{2.36}$$

Let $n \geq 2$ be an integer, $\{P_i\}_{i=0}^{n-1} \subset \mathcal{M}$ satisfy

$$P_i(X) = C, \quad i = 0, \ldots, n - 1, \tag{2.37}$$

$$\alpha_i \in (0, 1], \quad i = 1, \ldots, n - 1, \quad \{x_i\}_{i=0}^{n} \subset X, \quad \{\xi_i\}_{i=0}^{n-1} \subset X, $$

$$\|x_0\| \leq K_1, \tag{2.38}$$

$$\|x_1 - P_0 x_0\| \leq \delta_C \tag{2.39}$$

and that for all $i = 1, \ldots, n - 1$,

$$B_X(\xi_i, \delta_f) \cap \partial_{\epsilon/4} f(x_i) \neq \emptyset, \tag{2.40}$$

$$\|x_{i+1} - P_i(x_i \doteq \alpha_i\xi_i)\| \leq \delta_C. \tag{2.41}$$

Then

$$\min\{f(x_i) : i = 1, \ldots, n\} - \inf(f, C), \ f\left(\left(\sum_{j=1}^{n} \alpha_j\right)^{-1} \sum_{i=1}^{n} \alpha_i x_i\right) - \inf(f, C)$$

$$\leq \left(\sum_{j=1}^{n} \alpha_j\right)^{-1} \sum_{i=1}^{n} \alpha_i (f(x_i) - \inf(f, C))$$

$$\leq \left(\sum_{i=1}^{n} \alpha_i\right)^{-1} (2K_1 + 1)^2 + 3\epsilon/4 + 25(L_1 + \bar{L})^2 \left(\sum_{i=1}^{n} \alpha_i^2\right)\left(\sum_{i=1}^{n} \alpha_i\right)^{-1}$$

$$+ 2\left(\sum_{i=1}^{n} \alpha_i\right)^{-1} \delta_C(\bar{K} + K_1 + 5L_1 + 5\bar{L} + 2)n.$$

Let $n \geq 2$ be an integer and $A > 0$ be given. We are interested in an optimal choice of the step-sizes α_i, $i = 1, \ldots, n$ satisfying $\sum_{i=1}^{n} \alpha_i = A$ which minimizes the right-hand side of the final equation in the statement of Theorem 2.3. In order to meet this goal, we need to minimize the function

$$\phi(\alpha_1, \ldots, \alpha_n) = \sum_{i=1}^{n} \alpha_i^2$$

on the set $\{(\alpha_1, \ldots, \alpha_n) \in R^n : \ \alpha_i \geq 0, \ i = 1, \ldots, n, \ \sum_{i=1}^{n} \alpha_i = A\}$. By Lemma 2.3 of [75], the minimizer of ϕ is $\alpha_i = n^{-1}A$, $i = 1, \ldots, n$.

Theorem 2.3 implies the following result.

Theorem 2.4 *Assume that $K_1 \geq \bar{K} + 1$, $L_1 \geq \bar{L}$, $\delta_f, \delta_C \in (0, 1]$,*

$$C \subset B_X(0, K_1),$$

$$|f(z_1) - f(z_2)| \leq L_1 \|z_1 - z_2\| \text{ for all } z_1, z_2 \in B_X(0, K_1 + 2),$$

$$0 < \epsilon \leq 16(\bar{L} + L_1 + 1)$$

and that

$$\delta_f(\bar{K} + K_1 + 2 + 5L_1 + 5\bar{L}) \leq \epsilon/8.$$

Let $n \geq 2$ be an integer, $\{P_i\}_{i=0}^{n-1} \subset \mathcal{M}$ satisfy

$$P_i(X) = C, \ i = 0, \ldots, n - 1,$$

$$\alpha \in (0, 1], \ \{x_i\}_{i=0}^{n} \subset X, \ \{\xi_i\}_{i=0}^{n-1} \subset X, \ \|x_0\| \leq K_1, \ \|x_1 - P_0 x_0\| \leq \delta_C$$

and that for all $i = 1, \ldots, n - 1$,

$$B_X(\xi_i, \delta_f) \cap \partial_{\epsilon/4} f(x_i) \neq \emptyset,$$

$$\|x_{i+1} - P_i(x_i - \alpha \xi_i)\| \leq \delta_C.$$

Then

$$\min\{f(x_i) : i = 1, \ldots, n\} - \inf(f, C), \quad f\left(n^{-1} \sum_{i=1}^{n} x_i\right) - \inf(f, C)$$

$$\leq n^{-1} \sum_{i=1}^{n} f(x_i) - \inf(f, C)$$

$$\leq (n\alpha)^{-1}(2K_1 + 1)^2 + 3\epsilon/4 + 25(L_1 + \bar{L})^2 \alpha$$
$$+ 2\alpha^{-1} \delta_C (\bar{K} + K_1 + 5L_1 + 5\bar{L} + 2).$$

Now we can make the best choice of the step-size α in Theorem 2.4. Since n can be arbitrary large in view of Theorem 2.4, we need to minimize the function

$$25(L_1 + \bar{L})^2 \alpha + 2\alpha^{-1} \delta_C (\bar{K} + K_1 + 5L_1 + 5\bar{L} + 2), \quad \alpha > 0$$

which has a minimizer

$$\alpha = 5^{-1}(L_1 + \bar{L})^{-1}(2(\bar{K} + K_1 + 5L_1 + 5\bar{L} + 2))^{1/2} \delta_C^{1/2}.$$

With this choice of α, the right-hand side of the last equation in the statement of Theorem 2.4 is

$$n^{-1}(2K_1 + 1)^2 5(L_1 + \bar{L})(2(\bar{K} + K_1 + 5L_1 + 5\bar{L} + 2))^{-1/2} \delta_C^{-1/2}$$

$$+ 3\epsilon/4 + 10(L_1 + \bar{L})(2(\bar{K} + K_1 + 5L_1 + 5\bar{L} + 2))^{1/2} \delta_C^{1/2}.$$

Now we should make the best choice of n. It is clear that n should be at the same order as δ_C^{-1}. In this case, the right-hand side of the last equation in Theorem 2.4 does not exceed $c_1 \delta_C^{1/2} + 3\epsilon/4$. Clearly, ϵ depends on δ_f. In particular, we can choose

$$\epsilon = 8\delta_f(\bar{K} + K_1 + 5L_1 + 5\bar{L} + 2).$$

In the next theorem, the set C is not necessarily bounded, and the computational errors δ_f, δ_C and the step-size α are given. It is proved in Section 2.9.

Theorem 2.5 *Assume that $K_1 \geq \bar{K} + 2$, $L_1 \geq \bar{L}$, $\delta_f, \delta_C \in (0, 1]$, $\epsilon \in (0, 4]$,*

$$|f(z_1) - f(z_2)| \leq L_1 \|z_1 - z_2\| \text{ for all } z_1, z_2 \in B_X(0, 3K_1 + 1), \tag{2.42}$$

$$\alpha \in (0, 25^{-1}(L_1 + \bar{L})^{-2}] \tag{2.43}$$

and that

$$8\delta_f(3\bar{K} + K_1 + 2 + 5L_1 + 5\bar{L}) \leq \epsilon, \tag{2.44}$$

$$\delta_C(3\bar{K} + K_1 + 2 + 5L_1 + 5\bar{L}) \leq \alpha. \tag{2.45}$$

Let $n \geq 2$ be an integer, $\{P_i\}_{i=0}^{n-1} \subset \mathcal{M}$ satisfy

$$P_i(X) = C, \ i = 0, \ldots, n - 1, \tag{2.46}$$

$$\{x_i\}_{i=0}^n \subset X, \ \{\xi_i\}_{i=0}^{n-1} \subset X,$$

$$\|x_0\| \leq K_1, \tag{2.47}$$

$$\|x_1 - P_0 x_0\| \leq \delta_C \tag{2.48}$$

and that for all $i = 1, \ldots, n - 1$,

$$B_X(\xi_i, \delta_f) \cap \partial_{\epsilon/4} f(x_i) \neq \emptyset, \tag{2.49}$$

$$\|x_{i+1} - P_i(x_i - \alpha \xi_i)\| \leq \delta_C. \tag{2.50}$$

Then

$$\|x_i\| \leq 2\bar{K} + K_1 + 1, \ i = 1, \ldots, n$$

and

$$\min\{f(x_i) : t = 1, \ldots, n\} - \inf(f, C), \ f\left(n^{-1} \sum_{i=1}^n x_i\right) - \inf(f, C)$$

$$\leq n^{-1} \sum_{i=1}^n f(x_i) - \inf(f, C)$$

$$\leq (2n\alpha)^{-1}(K_1 + 1 + \bar{K})^2 + \epsilon/2 + 15(L_1 + \bar{L})^2\alpha$$
$$+ \alpha^{-1}\delta_C(3\bar{K} + K_1 + 5L_1 + 5\bar{L} + 2).$$

Now we can make the best choice of the step-size α in Theorem 2.5. Since n can be arbitrary large in view of Theorem 2.5, we need to minimize the function

$$15(L_1 + \bar{L})^2\alpha + \alpha^{-1}\delta_C(3\bar{K} + K_1 + 5L_1 + 5\bar{L} + 2), \ \alpha > 0$$

which has a minimizer

$$\alpha = (L_1 + \bar{L})^{-1}(15^{-1}(3\bar{K} + K_1 + 5L_1 + 5\bar{L} + 2))^{1/2}\delta_C^{1/2}.$$

Since α should satisfy (2.45), we obtain an additional condition on δ_C:

$$(3\bar{K} + K_1 + 5L_1 + 5\bar{L} + 2)\delta_C \leq 15 \cdot 25^{-2}(L_1 + \bar{L})^{-2}.$$

Together with the relations above Theorem 2.5 implies the following result.

Theorem 2.6 *Assume that $K_1 \geq \bar{K} + 2$, $L_1 \geq \bar{L}$, $\delta_f, \delta_C \in (0, 1]$, $\epsilon \in (0, 4]$,*

$$|f(z_1) - f(z_2)| \leq L_1\|z_1 - z_2\| \text{ for all } z_1, z_2 \in B_X(0, 3K_1 + 1),$$

$$\delta_f \leq 8^{-1}(3\bar{K} + K_1 + 2 + 5L_1 + 5\bar{L})\epsilon,$$

$$\delta_C \leq 25^{-2}15(L_1 + \bar{L})^{-2}(3\bar{K} + K_1 + 2 + 5L_1 + 5\bar{L})^{-1}$$

and that

$$\alpha = (L_1 + \bar{L})^{-1}(15^{-1}(3\bar{K} + K_1 + 5L_1 + 5\bar{L} + 2))^{1/2}\delta_C^{1/2}.$$

Let $n \geq 2$ be an integer, $\{P_i\}_{i=0}^{n-1} \subset \mathcal{M}$ satisfy

$$P_i(X) = C, \ i = 0, \ldots, n - 1,$$

$$\{x_i\}_{i=0}^{n} \subset X, \ \{\xi_i\}_{i=0}^{n-1} \subset X,$$

$$\|x_0\| \leq K_1,$$

$$\|x_1 - P_0 x_0\| \leq \delta_C$$

and that for all $i = 1, \ldots, n - 1$,

$$B_X(\xi_i, \delta_f) \cap \partial_{\epsilon/4} f(x_i) \neq \emptyset,$$

$$\|x_{i+1} - P_i(x_i - \alpha\xi_i)\| \leq \delta_C.$$

Then

$$\|x_i\| \leq 2\bar{K} + K_1 + 1, \ i = 1, \ldots, n$$

and

$$\min\{f(x_i) : t = 1, \ldots, n\} - \inf(f, C), \quad f\left(n^{-1}\sum_{i=1}^{n} x_i\right) - \inf(f, C)$$

$$\leq n^{-1}\sum_{i=1}^{n} f(x_i) - \inf(f, C)$$

$$\leq \epsilon/2 + 2^{-1}n^{-1}(K_1 + 1 + \bar{K})^2(L_1 + \bar{L})15^{1/2}(\delta_C(3\bar{K} + K_1 + 5L_1 + 5\bar{L} + 2))^{-1/}$$

$$+30(L_1 + \bar{L})(15^{-1}(3\bar{K} + K_1 + 5L_1 + 5\bar{L} + 2)\delta_C)^{1/2}.$$

Now we should make the best choice of n in Theorem 2.6. It is clear that n should be at the same order as δ_C^{-1}.

Theorem 2.6 implies the following result.

Theorem 2.7 *Assume that* $K_1 \geq \bar{K} + 2$, $L_1 \geq \bar{L}$, $\delta_f, \delta_C \in (0, 1]$,

$$|f(z_1) - f(z_2)| \leq L_1\|z_1 - z_2\| \text{ for all } z_1, z_2 \in B_X(0, 3K_1 + 1),$$

$$\delta_f \leq 2^{-1}(3\bar{K} + K_1 + 2 + 5L_1 + 5\bar{L})^{-1},$$

$$\delta_C \leq 25^{-2}15(L_1 + \bar{L})^{-2}(3\bar{K} + K_1 + 2 + 5L_1 + 5\bar{L})^{-1},$$

$$\epsilon = 8\delta_f(3\bar{K} + K_1 + 2 + 5L_1 + 5\bar{L})$$

and that

$$\alpha = (L_1 + \bar{L})^{-1}(15^{-1}(3\bar{K} + K_1 + 5L_1 + 5\bar{L} + 2))^{1/2}\delta_C^{1/2}.$$

Let $n \geq 2$ *be an integer,* $\{P_i\}_{i=0}^{n-1} \subset \mathcal{M}$ *satisfy*

$$P_i(X) = C, \quad i = 0, \ldots, n - 1,$$

$$\{x_i\}_{i=0}^{n} \subset X, \quad \{\xi_i\}_{i=0}^{n-1} \subset X,$$

$$\|x_0\| \leq K_1,$$

$$\|x_1 - P_0 x_0\| \leq \delta_C$$

and that for all $i = 1, \ldots, n - 1$,

$$B_X(\xi_i, \delta_f) \cap \partial_{\epsilon/4} f(x_i) \neq \emptyset,$$

$$\|x_{i+1} - P_i(x_i - \alpha\xi_i)\| \leq \delta_C.$$

Then

$$\|x_i\| \le 2\bar{K} + K_1 + 1, \ i = 1, \ldots, n$$

and

$$\min\{f(x_i) : \ i = 1, \ldots, n\} - \inf(f, C), \ f\left(n^{-1} \sum_{i=1}^{n} x_i\right) - \inf(f, C)$$

$$\le n^{-1} \sum_{i=1}^{n} f(x_i) - \inf(f, C)$$

$$\le 4\delta_f (3\bar{K} + K_1 + 2 + 5L_1 + 5\bar{L})$$
$$+ 2^{-1} n^{-1} (K_1 + 1 + \bar{K})^2 (L_1 + \bar{L}) 15^{1/2} (\delta_C (3\bar{K} + K_1 + 5L_1 + 5\bar{L} + 2))^{-1/2}$$
$$+ 30(L_1 + \bar{L})(15^{-1} (3\bar{K} + K_1 + 5L_1 + 5\bar{L} + 2)\delta_C)^{1/2}.$$

Now we should make the best choice of n in Theorem 2.7. It is clear that n should be at the same order as δ_C^{-1}. In this case, the right-hand side of the last equation in Theorem 2.7 does not exceed $c_1 \delta_f + c_2 \delta_C^{1/2}$ where $c_1, c_2 > 0$ are constants.

In the previous theorems, we deal with the projected subgradient method: at every iterative step i of the algorithm, ξ_i is an approximation of a subgradient v. In the following theorems, we study the projected normalized subgradient method: at every iterative step i of the algorithm, ξ_i is an approximation of $\|v\|^{-1} v$, where v is a subgradient.

In the next result, the set C is bounded, and the computational errors δ_f, δ_C and ϵ are given. It is proved in Section 2.10.

Theorem 2.8 *Assume that $K_1 \ge \bar{K} + 1$,*

$$C \subset B_X(0, K_1), \tag{2.51}$$

$\delta_f, \delta_C \in (0, 1),$

$$0 < \epsilon \le 16\bar{L}, \tag{2.52}$$

$$8\bar{L}\delta_f (\bar{K} + K_1 + 1) \le \epsilon, \tag{2.53}$$

$$(32\bar{L})^2 \delta_C (\bar{K} + K_1 + 3) \le \epsilon^2 \tag{2.54}$$

and that

$$n = \lfloor 2(16\bar{L})^2 (1 + K_1 + \bar{K})^2 \epsilon^{-2} \rfloor + 3. \tag{2.55}$$

Let $\{P_i\}_{i=0}^{n-1} \subset \mathcal{M}$ satisfy

$$P_i(X) = C, \quad i = 0, \ldots, n-1, \tag{2.56}$$

$\{x_i\}_{i=0}^n \subset X$, $\{\xi_i\}_{i=1}^{n-1} \subset X$,

$$\{\alpha_i\}_{i=1}^{n-1} \subset [(32\bar{L})^{-1}\epsilon, (16\bar{L})^{-1}\epsilon], \tag{2.57}$$

$$\|x_0\| \leq K_1, \tag{2.58}$$

$$\|x_1 - P_0 x_0\| \leq \delta_C \tag{2.59}$$

and that for all $i = 1, \ldots, n-1$,

$$B_X(\xi_i, \delta_f) \cap \{\|v\|^{-1} v : v \in \partial_{\epsilon/4} f(x_i) \setminus \{0\}\} \neq \emptyset, \tag{2.60}$$

$$\|x_{i+1} - P_i(x_i - \alpha_i \xi_i)\| \leq \delta_C. \tag{2.61}$$

Then there exists $j \in \{1, \ldots, n\}$ such that

$$f(x_j) \leq \inf(f, C) + \epsilon$$

and if $i \in \{1, \ldots, n\} \setminus \{j\}$, then

$$f(x_i) > \inf(f, C) + \epsilon$$

and

$$\partial_{\epsilon/4} f(x_i) \subset \{v \in X : \|v\| \geq (3/4)\epsilon(K_1 + \bar{K})^{-1}\}.$$

Theorem 2.8 implies the following result.

Theorem 2.9 *Assume that $K_1 \geq \bar{K} + 1$,*

$$C \subset B_X(0, K_1),$$

$\delta_f, \delta_C \in (0, 1)$,

$$\delta_f(\bar{K} + K_1 + 1) \leq 2,$$

$$\delta_C(\bar{K} + K_1 + 3) \leq 4^{-1},$$

$$\epsilon = \max\{8\bar{L}\delta_f(\bar{K} + K_1 + 1), \ 32\bar{L}(\delta_C(\bar{K} + K_1 + 3))^{1/2}\}$$

and that

$$n = \lfloor 2(16\bar{L})^2 (K_1 + \bar{K})^2 \epsilon^{-2} \rfloor + 2.$$

Let $\{P_i\}_{i=0}^{n-1} \subset \mathcal{M}$ satisfy

$$P_i(X) = C, \ i = 0, \ldots, n-1,$$

$\{x_i\}_{i=0}^{n} \subset X, \ \{\xi_i\}_{i=1}^{n-1} \subset X,$

$$\{\alpha_i\}_{i=1}^{n-1} \subset [(32\bar{L})^{-1}\epsilon, (16\bar{L})^{-1}\epsilon],$$

$$\|x_0\| \le K_1,$$

$$\|x_1 - P_0 x_0\| \le \delta_C$$

and that for $i = 1, \ldots, n-1$,

$$B_X(\xi_i, \delta_f) \cap \{\|v\|^{-1} v : v \in \partial_{\epsilon/4} f(x_i) \setminus \{0\}\} \neq \emptyset,$$

$$\|x_{i+1} - P_i(x_i - \alpha_i \xi_i)\| \le \delta_C.$$

Then there exists $j \in \{1, \ldots, n\}$ such that

$$f(x_j) \le \inf(f, C) + \epsilon$$

and if $i \in \{1, \ldots, n\} \setminus \{j\}$, then

$$f(x_i) > \inf(f, C) + \epsilon$$

and

$$\partial_{\epsilon/4} f(x_i) \subset \{v \in X : \|v\| \ge (3/4)\epsilon(K_1 + \bar{K})^{-1}\}.$$

In the next theorem, the set C is not necessarily bounded. It is proved in Section 2.11.

Theorem 2.10 *Assume that* $K_1 \ge \bar{K} + 2$, $\delta_f, \delta_C \in (0, 1]$,

$$4\bar{L}\delta_f(3\bar{K} + K_1 + 2) \le 1, \tag{2.62}$$

$$(8\bar{L})^2 \delta_C(3\bar{K} + K_1 + 4) \le 1, \tag{2.63}$$

$$\epsilon = \max\{8\bar{L}\delta_f(3\bar{K} + K_1 + 2), \ 32\bar{L}(\delta_C(3\bar{K} + K_1 + 4))^{1/2}\} \tag{2.64}$$

and that

$$n = \lfloor 2(16\bar{L})^2 (1 + K_1 + \bar{K})^2 \epsilon^{-2} \rfloor + 2. \tag{2.65}$$

Let $\{P_i\}_{i=0}^{n-1} \subset \mathcal{M}$ satisfy

$$P_i(X) = C, \ i = 0, \ldots, n - 1, \tag{2.66}$$

$\{x_i\}_{i=0}^{n} \subset X, \ \{\xi_i\}_{i=1}^{n-1} \subset X,$

$$\{\alpha_i\}_{i=1}^{n-1} \subset [(32\bar{L})^{-1}\epsilon, (16\bar{L})^{-1}\epsilon] \subset (0, 1], \tag{2.67}$$

$$\|x_0\| \le K_1, \tag{2.68}$$

$$\|x_1 - P_0 x_0\| \le \delta_C \tag{2.69}$$

and that for $i = 1, \ldots, n - 1$,

$$B_X(\xi_i, \delta_f) \cap \{\|v\|^{-1} v : \ v \in \partial_{\epsilon/4} f(x_i) \setminus \{0\}\} \ne \emptyset, \tag{2.70}$$

$$\|x_{i+1} - P_i(x_i - \alpha_i \xi_i)\| \le \delta_C. \tag{2.71}$$

Then there exists $j \in \{1, \ldots, n\}$ such that

$$f(x_j) \le \inf(f, C) + \epsilon$$

and if $i \in \{1, \ldots, n\} \setminus \{j\}$, then

$$f(x_i) > \inf(f, C) + \epsilon$$

and

$$\partial_{\epsilon/4} f(x_i) \subset \{v \in X : \ \|v\| \ge (3/4)\epsilon(K_1 + 3\bar{K} + 1)^{-1}\}.$$

All the results of this section are new.

2.3 Convergence to the Solution Set

We use the notation and definitions introduced in Section 2.1 and suppose that all the assumptions made there hold. We continue to study the minimization problem

$$f(x) \to \min, \ x \in C.$$

In Section 2.2, we obtain an approximate solution x which is close to the set C such that $f(x)$ is closed to $\inf(f, C)$. In this section, we obtain an approximate solution x which is close to the set C_{min}.

We also suppose that the following assumption holds.

(A1) For every positive number ϵ, there exists $\delta > 0$ such that if a point $x \in C$ satisfies the inequality $f(x) \leq \inf(f, C) + \delta$, then $d(x, C_{min}) \leq \epsilon$.

(It is clear that (A1) holds if the space X is finite-dimensional.)
For every number $\epsilon \in (0, \infty)$, let

$$\phi(\epsilon) = \sup\{\delta \in (0, 1] : \text{ if } x \in C \text{ satisfies } f(x) \leq \inf(f, C) + \delta,$$

$$\text{then } d(x, C_{min}) \leq \min\{1, \epsilon\}\}. \tag{2.72}$$

In view of (A1), $\phi(\epsilon)$ is well-defined for every positive number ϵ.

In the following result, the step-sizes converge to zero, and their sum is infinity. It is proved in Section 2.13.

Theorem 2.11 *Let* $\{\alpha_i\}_{i=0}^{\infty} \subset (0, 1]$ *satisfy*

$$\lim_{i \to \infty} \alpha_i = 0, \quad \sum_{i=1}^{\infty} \alpha_i = \infty \tag{2.73}$$

and let $M, \epsilon > 0$. *Then there exist a natural number* n_0 *and* $\delta > 0$ *such that the following assertion holds.*
Assume that an integer $n \geq n_0$, $\{P_i\}_{i=0}^{n-1} \subset \mathcal{M}$,

$$P_i(X) = C, \ i = 0, \dots, n - 1, \tag{2.74}$$

$$\{x_i\}_{i=0}^{n} \subset X, \ \|x_0\| \leq M, \tag{2.75}$$

$$v_i \in \partial_\delta f(x_i) \setminus \{0\}, \ i = 0, 1, \dots, n - 1, \tag{2.76}$$

$$\{\eta_i\}_{i=0}^{n-1}, \ \{\xi_i\}_{i=0}^{n-1} \subset B_X(0, \delta),$$

and that for $i = 0, \dots, n - 1$,

$$x_{i+1} = P_i(x_i - \alpha_i \|v_i\|^{-1} v_i - \alpha_i \xi_i) - \alpha_i \eta_i.$$

Then the inequality $d(x_i, C_{min}) \leq \epsilon$ *holds for all integers* i *satisfying* $n_0 \leq i \leq n$.

In the following theorem, the step-sizes are bounded from below by a sufficiently small positive constant. It is proved in Section 2.14.

Theorem 2.12 *Let* $M, \epsilon > 0$. *Then there exists* $\beta_0 \in (0, 1)$ *such that for each* $\beta_1 \in (0, \beta_0)$, *there exist a natural number* n_0 *and* $\delta > 0$ *such that the following assertion holds.*
Assume that an integer $n \geq n_0$, $\{P_i\}_{i=0}^{n-1} \subset \mathcal{M}$,

$$P_i(X) = C, \ i = 0, \ldots, n - 1,$$

$$\{x_i\}_{i=0}^n \subset X, \ \|x_0\| \leq M, \tag{2.77}$$

$$v_i \in \partial_\delta f(x_i) \setminus \{0\}, \ i = 0, 1, \ldots, n - 1, \tag{2.78}$$

$$\{\alpha_i\}_{i=0}^{n-1} \subset [\beta_1, \beta_0], \tag{2.79}$$

$$\{\eta_i\}_{i=0}^{n-1}, \ \{\xi_i\}_{i=0}^{n-1} \subset B_X(0, \delta) \tag{2.80}$$

and that for $i = 0, \ldots, n - 1$,

$$x_{i+1} = P_i(x_i - \alpha_i \|v_i\|^{-1} v_i - \alpha_i \xi_i) - \eta_i. \tag{2.81}$$

Then the inequality $d(x_i, C_{min}) \leq \epsilon$ holds for all integers i satisfying $n_0 \leq i \leq n$.

In the previous two theorems, we deal with the projected normalized subgradient method: at every iterative step i of the algorithm, ξ_i is an approximation of $\|v\|^{-1} v$, where v is a subgradient. In the following three theorems, we study the projected subgradient method: at every iterative step i of the algorithm, ξ_i is an approximation of a subgradient v. In the next theorem, the set C is bounded, and the step-sizes converge to zero, and their sum is infinity. It is proved in Section 2.15.

Theorem 2.13 Let $\{\alpha_i\}_{i=0}^\infty \subset (0, 1]$ satisfy

$$\lim_{i \to \infty} \alpha_i = 0, \ \sum_{i=1}^\infty \alpha_i = \infty, \tag{2.82}$$

$$C \subset B_X(0, M) \tag{2.83}$$

and let $M, \epsilon > 0$. Then there exist a natural number n_0 and $\delta > 0$ such that the following assertion holds.

Assume that an integer $n \geq n_0$, $\{P_i\}_{i=0}^{n-1} \subset \mathcal{M}$,

$$P_i(X) = C, \ i = 0, \ldots, n - 1,$$

$$\{x_i\}_{i=0}^n \subset X, \ \|x_0\| \leq M, \tag{2.84}$$

$\{\xi_i\}_{i=0}^{n-1} \subset X$, for all $i = 0, \ldots, n - 1$,

$$B_X(\xi_i, \delta) \cap \partial_\delta f(x_i) \neq \emptyset, \tag{2.85}$$

$$\|x_{i+1} - P_i(x_i - \alpha_i \xi_i)\| \leq \delta_i. \tag{2.86}$$

Then the inequality $d(x_i, C_{min}) \leq \epsilon$ holds for all integers i satisfying $n_0 \leq i \leq n$.

In the next theorem, the set C is not necessarily bounded, but the step-sizes converge to zero, and their sum is infinity. It is proved in Section 2.16.

Theorem 2.14 *Let*

$$M > 2\bar{K} + 1, \ \epsilon \in (0, 1), \ L_0 > \bar{L},$$

$$|f(z_1) - f(z_2)| \le L_0 \|z_1 - z_2\| \text{ for all } z_1, z_2 \in B_X(0, 2M + 4). \tag{2.87}$$

and let $\{\alpha_i\}_{i=0}^{\infty} \subset (0, 1]$ *satisfy*

$$\lim_{i \to \infty} \alpha_i = 0, \ \sum_{i=1}^{\infty} \alpha_i = \infty,$$

$$\alpha_t \le 25^{-2}(\bar{L} + L_0 + 2)^{-2}, \ t = 0, 1, \dots. \tag{2.88}$$

Then there exist a natural number n_0 and $\delta > 0$ such that the following assertion holds.
Assume that an integer $n \ge n_0$, $\{P_i\}_{i=0}^{n-1} \subset \mathcal{M}$,

$$P_i(X) = C, \ i = 0, \dots, n - 1,$$

$$\{x_i\}_{i=0}^{n} \subset X, \ \|x_0\| \le M,$$

$\{\xi_i\}_{i=0}^{n-1} \subset X,$ *for all* $i = 0, \dots, n - 1,$

$$B_X(\xi_i, \delta) \cap \partial_\delta f(x_i) \ne \emptyset,$$

$$\|x_{i+1} - P_i(x_i - \alpha_i \xi_i)\| \le \alpha_i \delta. \tag{2.89}$$

Then the inequality $d(x_i, C_{min}) \le \epsilon$ holds for all integers i satisfying $n_0 \le i \le n$.

In the next theorem, the set C is not necessarily bounded, while the step-sizes are bounded from below by a positive constant. It is proved in Section 2.17.

Theorem 2.15 *Let $M, \epsilon > 0$. Then there exists $\beta_0 \in (0, 1)$ such that for each $\beta_1 \in (0, \beta_0)$, there exist a natural number n_0 and $\delta > 0$ such that the following assertion holds.*
Assume that an integer $n \ge n_0$, $\{P_i\}_{i=0}^{n-1} \subset \mathcal{M}$,

$$P_i(X) = C, \ i = 0, \dots, n - 1,$$

$$\{x_i\}_{i=0}^{n} \subset X, \ \|x_0\| \le M, \tag{2.90}$$

$$\{\alpha_i\}_{i=0}^{n-1} \subset [\beta_1, \beta_0], \tag{2.91}$$

$\{\xi_i\}_{i=0}^{n-1} \subset X$, for all $i = 0, \ldots, n-1$,

$$B_X(\xi_i, \delta) \cap \partial_\delta f(x_i) \neq \emptyset, \qquad (2.92)$$

$$\|x_{i+1} - P_i(x_i - \alpha_i \xi_i)\| \leq \delta. \qquad (2.93)$$

Then the inequality $d(x_i, C_{min}) \leq \epsilon$ holds for all integers i satisfying $n_0 \leq i \leq n$.

Theorems 2.13–2.15 are new. In the case with $P_i = P_0$, $i = 0, \ldots, n-1$, Theorems 2.11 and 2.12 were obtained in [71].

2.4 Superiorization

In this section, we present three results on the projected subgradient in the case when step-sizes are summable. Such results are of interest in the study of superiorization and perturbation resilience of algorithms. See [9, 13, 25, 29] and the references mentioned therein.

In the first two theorems at the iterative step i of the algorithm ξ_i is an approximation of a subgradient v, while in the third theorem ξ_i is an approximation of $\|v\|^{-1}v$, where v is a subgradient. In the first theorem, the set C is bounded. It is proved in Section 2.18.

Theorem 2.16 Let $K_1 \geq \bar{K} + 1$, $L_1 \geq \bar{L}$, $\{\alpha_i\}_{i=0}^\infty \subset (0, 1]$ satisfy

$$\sum_{i=1}^\infty \alpha_i < \infty,$$

$$C \subset B_X(0, K_1) \qquad (2.94)$$

and let

$$|f(z_1) - f(z_2)| \leq L_1 \|z_1 - z_2\| \text{ for all } z_1, z_2 \in B_X(0, K_1 + 2). \qquad (2.95)$$

Assume that $\{P_i\}_{i=0}^\infty \subset \mathcal{M}$,

$$P_i(X) = C, \ i = 0, \ldots, \qquad (2.96)$$

$\{x_i\}_{i=0}^\infty \subset X$, $\{\xi_i\}_{i=0}^\infty \subset X$,

$$\|x_0\| \leq K_1, \qquad (2.97)$$

for all integers $i \geq 0$,

$$\xi_i \in \partial f(x_i), \tag{2.98}$$

$$x_{i+1} = P_i(x_i - \alpha_i \xi_i). \tag{2.99}$$

Then there exists $x_* = \lim_{i \to \infty} x_i$, and at least one of the following properties holds:

$$x_* \in C_{min};$$

there exists an integer $n_0 \geq 1$ and $\epsilon_0 > 0$ such that for each integer $t \geq n_0$ and each $z \in C_{min}$,

$$\|x_{t+1} - z\|^2 \leq \|x_t - z\|^2 - \alpha_t \epsilon.$$

In the following two theorems, the set C is not necessarily bounded. The first of them is proved in Section 2.19.

Theorem 2.17 Let $K_1 \geq \bar{K} + 1$, $L_1 \geq \bar{L}$, $\{\alpha_i\}_{i=0}^\infty \subset (0, 1]$,

$$|f(z_1) - f(z_2)| \leq L_1 \|z_1 - z_2\| \text{ for all } z_1, z_2 \in B_X(0, 3K_1 + 2). \tag{2.100}$$

Assume that $\{P_i\}_{i=0}^\infty \subset \mathcal{M}$,

$$P_i(X) = C, \ i = 0, 1, \ldots, \tag{2.101}$$

$\{x_i\}_{i=0}^\infty \subset X$, $\{\xi_i\}_{i=0}^\infty \subset X$,

$$\|x_0\| \leq K_1,$$

for all integers $i \geq 0$,

$$\alpha_i \in (0, 100(L_1 + \bar{L})^{-2}], \tag{2.102}$$

$$\xi_i \in \partial f(x_i), \tag{2.103}$$

$$x_{i+1} = P_i(x_i - \alpha_i \xi_i). \tag{2.104}$$

Then there exists $x_* = \lim_{i \to \infty} x_i$, and at least one of the following properties holds:

$$x_* \in C_{min};$$

there exists an integer $n_0 \geq 1$ and $\epsilon_0 > 0$ such that for each integer $t \geq n_0$ and each $z \in C_{min}$,

$$\|x_{t+1} - z\|^2 \le \|x_t - z\|^2 - \alpha_t \epsilon_0.$$

The next result is proved in Section 2.20.

Theorem 2.18 *Let $K_1 \ge \bar{K} + 1$, $\{\alpha_i\}_{i=0}^{\infty} \subset (0, 1]$ satisfy*

$$\sum_{i=1}^{\infty} \alpha_i < \infty,$$

Assume that $\{P_i\}_{i=0}^{\infty} \subset \mathcal{M}$,

$$P_i(X) = C, \ i = 0, 1, \ldots, \tag{2.105}$$

$\{x_i\}_{i=0}^{\infty} \subset X$, $\{\xi_i\}_{i=0}^{\infty} \subset X \setminus \{0\}$,

$$\|x_0\| \le K_1, \tag{2.106}$$

for all integers $i \ge 0$,

$$\xi_i \in \partial f(x_i), \tag{2.107}$$

$$x_{i+1} = P_i(x_i - \alpha_i \|\xi_i\|^{-1} \xi_i). \tag{2.108}$$

Then there exists $x_* = \lim_{i \to \infty} x_i$, and at least one of the following properties holds:

$$x_* \in C_{min};$$

there exists an integer $n_0 \ge 1$ and $\epsilon_0 > 0$ such that for each integer $t \ge n_0$ and each $z \in C_{min}$,

$$\|x_{t+1} - z\|^2 \le \|x_t - z\|^2 - \alpha_t \epsilon.$$

2.5 Auxiliary Results for Theorems 2.1–2.10

Let

$$P_C \in \mathcal{M} \tag{2.109}$$

be an arbitrary element of the space $\mathcal{M}$.

Lemma 2.19 *Let $K_0, L_0, r > 0$,*

$$|f(z_1) - f(z_2)| \leq L_0 \|z_1 - z_2\| \text{ for all } z_1, z_2 \in B_X(0, K_0 + 1),$$ (2.110)

$$x \in B_X(0, K_0),$$ (2.111)

$$v \in \partial f_r(x).$$ (2.112)

Then $\|v\| \leq L_0 + r$.

Proof By (2.112), for all $u \in X$,

$$f(u) - f(x) \geq \langle v, u - x \rangle - r.$$ (2.113)

In view of (2.113), for all $\xi \in B_X(0, 1)$,

$$r + f(x + \xi) - f(x) \geq \langle v, \xi \rangle.$$

Together with (2.110) and (2.111), this implies that

$$\langle v, \xi \rangle \leq r + L_0 \|\xi\| \leq r + L_0.$$

This implies that $\|v\| \leq L_0 + 2$. Lemma 2.19 is proved.

Lemma 2.20 *Assume that* $\epsilon > 0$, $x \in X$, $y \in X$,

$$f(x) > \inf(f, C) + \epsilon,$$ (2.114)

$$f(y) \leq \inf(f, C) + \epsilon/4,$$ (2.115)

$$v \in \partial_{\epsilon/4} f(x).$$ (2.116)

Then $\langle v, y - x \rangle \leq -\epsilon/2$.

Proof In view of (2.116), for all $u \in X$,

$$f(u) - f(x) \geq \langle v, u - x \rangle - \epsilon/4.$$ (2.117)

By (2.115),

$$-(3/4)\epsilon \geq f(y) - f(x) \geq \langle v, y - x \rangle - \epsilon/4.$$

The inequality above implies that

$$\langle v, y - x \rangle \leq -\epsilon/2.$$

This completes the proof of Lemma 2.20.

Lemma 2.21 *Let*

$$\bar{x} \in C_{min}, \tag{2.118}$$

$K_0 > 0, L_0 > 0,$

$$|f(z_1) - f(z_2)| \leq L_0 \|z_1 - z_2\| \text{ for all } z_1, z_2 \in B_X(0, K_0 + 1), \tag{2.119}$$

$$\epsilon \in (0, 16(L_0 + \bar{L})], \tag{2.120}$$

$\alpha \in (0, 1], \delta_f, \delta_C \in (0, 1]$ *satisfy*

$$\delta_f(K_0 + \bar{K} + 5L_0 + 5\bar{L} + 1) \leq \epsilon/4, \tag{2.121}$$

let a point $x \in X$ satisfy

$$\|x\| \leq K_0, \quad f(x) > \inf(f, C) + \epsilon, \tag{2.122}$$

$$v \in \partial_{\epsilon/4} f(x), \tag{2.123}$$

$$\xi \in B_X(v, \delta_f) \tag{2.124}$$

and let

$$y \in B_X(P_C(x - \alpha\xi), \delta_C). \tag{2.125}$$

Then

$$\|y - \bar{x}\|^2 \leq \|x - \bar{x}\|^2 - 2^{-1}\alpha\epsilon + \delta_C^2$$
$$+ 2\delta_C(K_0 + \bar{K} + 5L_0 + 5\bar{L}) + 25(L_0 + \bar{L})^2\alpha^2$$

Proof Lemma 2.19, (2.119)–(2.121), and (2.123) imply that

$$\|v\| \leq 5L_0 + 4\bar{L}.$$

Lemma 2.20, (2.118), (2.122), and (2.123) imply that

$$\langle v, \bar{x} - x \rangle \leq -\epsilon/2. \tag{2.126}$$

Set

$$y_0 = x - \alpha\xi. \tag{2.127}$$

It follows from (2.7), (2.8), (2.118), (2.120), (2.121), (2.124), and (2.126) that

$$
\begin{aligned}
\|y_0 - \bar{x}\|^2 &= \|x - \alpha\xi - \bar{x}\|^2 \\
&= \|x - \alpha v + (\alpha v - \alpha\xi) - \bar{x}\|^2 \\
&= \|x - \alpha v - \bar{x}\|^2 + \alpha^2\|v - \xi\|^2 + 2\alpha\langle v - \xi, x - \alpha v - \bar{x}\rangle \\
&\leq \|x - \alpha v - \bar{x}\|^2 + \alpha^2\delta_f^2 + 2\alpha\delta_f\|x - \alpha v - \bar{x}\| \\
&\leq \|x - \alpha v - \bar{x}\|^2 + \alpha^2\delta_f^2 + 2\alpha\delta_f(K_0 + \bar{K} + \alpha(5L_0 + 4\bar{L})) \\
&\leq \|x - \bar{x}\|^2 - 2\alpha\langle x - \bar{x}, v\rangle + \alpha^2(5L_0 + 4\bar{L})^2 \\
&\quad + \alpha^2\delta_f^2 + 2\alpha\delta_f(K_0 + \bar{K} + \alpha(5L_0 + 4\bar{L})) \\
&\leq \|x - \bar{x}\|^2 - 2\alpha(\epsilon/2) + \alpha^2(5L_0 + 4\bar{L})^2 \\
&\quad + \alpha^2\delta_f^2 + 2\alpha\delta_f(K_0 + \bar{K} + \alpha(5L_0 + 5\bar{L})) \\
&\leq \|x - \bar{x}\|^2 - \alpha\epsilon + 25\alpha^2(L_0 + \bar{L})^2 \\
&\quad + 2\alpha\delta_f(K_0 + \bar{K} + 5\alpha(L_0 + \bar{L}) + 1) \\
&\leq \|x - \bar{x}\|^2 - \alpha\epsilon + 25\alpha^2(L_0 + \bar{L})^2 + \epsilon\alpha/2 \\
&\leq \|x - \bar{x}\|^2 - \alpha\epsilon/2 + 25\alpha^2(L_0 + \bar{L})^2.
\end{aligned}
\tag{2.128}
$$

In view of (2.128),

$$
\|y_0 - \bar{x}\|^2 \leq \|x - \bar{x}\|^2 + 25(L_0 + \bar{L})^2.
\tag{2.129}
$$

Relations (2.5), (2.7), (2.8), (2.118), (2.121), and (2.129) imply that

$$
\|y_0 - \bar{x}\| \leq K_0 + \bar{K} + 5(L_0 + \bar{L}).
\tag{2.130}
$$

By (2.109), (2.110), (2.118), (2.125), (2.127), (2.128), and (2.130),

$$
\begin{aligned}
\|y - \bar{x}\|^2 &= \|y - P_C(x - \alpha\xi) + P_C(x - \alpha\xi) - \bar{x}\|^2 \\
&\leq \|y - P_C(x - \alpha\xi)\|^2 + 2\|y - P_C(x - \alpha\xi)\|\|P_C(x - \alpha\xi) - \bar{x}\| \\
&\quad + \|P_C(x - \alpha\xi) - \bar{x}\|^2 \\
&\leq \delta_C^2 + 2\delta_C(K_0 + \bar{K} + 5L_0 + 5\bar{L}) + \|y_0 - \bar{x}\|^2 \\
&\leq \delta_C^2 + 2\delta_C(K_0 + \bar{K} + 5L_0 + 5\bar{L}) + \|x - \bar{x}\|^2 - \alpha\epsilon/2 + 25\alpha^2(L_0 + \bar{L})^2.
\end{aligned}
$$

Lemma 2.21 is proved.

Lemma 2.21 implies the following result.

Lemma 2.22 Let $K_0 > 0$, $L_0 > 0$,

$$|f(z_1) - f(z_2)| \leq L_0 \|z_1 - z_2\| \text{ for all } z_1, z_2 \in B_X(0, K_0 + 1),$$

$$\epsilon \in (0, 16(L_0 + \bar{L})],$$

$\alpha \in (0, 1]$, $\delta_f, \delta_C \in (0, 1]$ *satisfy*

$$\delta_f(K_0 + \bar{K} + 5L_0 + 5\bar{L} + 1) \leq \epsilon/4,$$

let a point $x \in X$ *satisfy*

$$\|x\| \leq K_0, \ f(x) > \inf(f, C) + \epsilon,$$

$$v \in \partial_{\epsilon/4} f(x),$$

$$\xi \in B_X(v, \delta_f),$$

and let

$$y \in B_X(P_C(x - \alpha\xi), \delta_C).$$

Then

$$d(y, C_{min})^2 \leq d(x, C_{min})^2 - 2^{-1}\alpha\epsilon + \delta_C^2$$
$$+ 2\delta_C(K_0 + \bar{K} + 5L_0 + 5\bar{L}) + 25(L_0 + \bar{L})^2\alpha^2.$$

Applying Lemma 2.22 with $\epsilon = 4$, we obtain the following result.

Lemma 2.23 *Let* $K_0 > 0$, $L_0 > 0$,

$$|f(z_1) - f(z_2)| \leq L_0 \|z_1 - z_2\| \text{ for all } z_1, z_2 \in B_X(0, K_0 + 1),$$

$\alpha \in (0, 1]$, $\delta_f, \delta_C \in (0, 1]$ *satisfy*

$$\delta_f(K_0 + \bar{K} + 5L_0 + 5\bar{L} + 1) \leq 1,$$

let a point $x \in X$ *satisfy*

$$\|x\| \leq K_0, \ f(x) > \inf(f, C) + 4,$$

$$v \in \partial_1 f(x),$$

$$\xi \in B_X(v, \delta_f),$$

and let

$$y \in B_X(P_C(x - \alpha\xi), \delta_C).$$

Then

$$d(y, C_{min})^2 \leq d(x, C_{min})^2 - 2\alpha + 2\delta_C(K_0 + \bar{K} + 5L_0 + 5\bar{L} + 1) + 25(L_0 + \bar{L})^2\alpha^2.$$

Recall that

$$P_C \in \mathcal{M} \tag{2.131}$$

be an arbitrary element of the space $\mathcal{M}$.

Lemma 2.24 *Let*

$$\bar{x} \in C_{min}, \tag{2.132}$$

$K_0 > 0,$

$$\epsilon \in (0, 16\bar{L}], \tag{2.133}$$

$\alpha \in (0, 1], \delta_f, \delta_C \in (0, 1]$ *satisfy*

$$\delta_f(K_0 + \bar{K} + 1) \leq (8\bar{L})^{-1}\epsilon, \tag{2.134}$$

let a point $x \in X$ satisfy

$$\|x\| \leq K_0, \tag{2.135}$$

$$f(x) > \inf(f, C) + \epsilon, \tag{2.136}$$

$$v \in \partial_{\epsilon/4} f(x). \tag{2.137}$$

Then

$$\|v\| \geq 3(4K_0 + 4\bar{K})^{-1}\epsilon \tag{2.138}$$

and for each

$$\xi \in B_X(\|v\|^{-1}v, \delta_f), \tag{2.139}$$

and each

$$y \in B_X(P_C(x - \alpha\xi), \delta_C) \tag{2.140}$$

the following inequality holds:

$$\|y - \bar{x}\|^2 \leq \|x - \bar{x}\|^2 - (4\bar{L})^{-1}\alpha\epsilon + 2\alpha^2 + \delta_C^2 + 2\delta_C(K_0 + \bar{K} + 2).$$

Proof In view of (2.5), (2.7)–(2.9), (2.132), and (2.133) for every point

$$z \in B_X(\bar{x}, 4^{-1}\epsilon\bar{L}^{-1}), \tag{2.141}$$

we have

$$f(z) \leq f(\bar{x}) + \bar{L}\|z - \bar{x}\| \leq f(\bar{x}) + \epsilon/4 = \inf(f, C) + \epsilon/4. \tag{2.142}$$

By (2.132), (2.136), and (2.137),

$$-\epsilon > f(\bar{x}) - f(x) \geq \langle v, \bar{x} - x \rangle > -\epsilon/4,$$

, and combined with (2.7), (2.8), (2.132), and (2.135), this implies that

$$\langle v, \bar{x} - x \rangle \leq -(3/4)\epsilon,$$

$$(3/4)\epsilon \leq \langle v, x - \bar{x} \rangle \leq \|v\|\|x - \bar{x}\| \leq \|v\|(K_0 + \bar{K}).$$

Therefore (2.138) is true.

Let $\xi, y \in X$ satisfy (2.139) and (2.140). Lemma 2.20, (2.136), (2.137), (2.141), and (2.142) imply that for every point

$$z \in B_X(\bar{x}, 4^{-1}\epsilon\bar{L}^{-1}),$$

we have

$$\langle v, z - x \rangle \leq -\epsilon/2.$$

Combined with (2.138), the inequality above implies that

$$\langle \|v\|^{-1}v, z - x \rangle < 0 \text{ for all } z \in B_X(\bar{x}, (4\bar{L})^{-1}\epsilon). \tag{2.143}$$

Set

$$\tilde{z} = \bar{x} + 4^{-1}\bar{L}^{-1}\epsilon\|v\|^{-1}v. \tag{2.144}$$

It is easy to see that

$$\tilde{z} \in B_X(\bar{x}, 4^{-1}\bar{L}^{-1}\epsilon). \tag{2.145}$$

Relations (2.143)–(2.145) imply that

$$0 > \langle \|v\|^{-1}v, \tilde{z} - x \rangle = \langle \|v\|^{-1}v, \bar{x} + 4^{-1}\bar{L}^{-1}\epsilon\|v\|^{-1}v - x \rangle. \qquad (2.146)$$

By (2.146),

$$\langle \|v\|^{-1}v, \bar{x} - x \rangle < -4^{-1}\bar{L}^{-1}\epsilon. \qquad (2.147)$$

Set

$$y_0 = x - \alpha\xi. \qquad (2.148)$$

It follows from (2.8), (2.131), (2.132), (2.134), (2.135), (2.139), and (2.148) that

$$
\begin{aligned}
\|y_0 - \bar{x}\|^2 &= \|x - \alpha\xi - \bar{x}\|^2 \\
&= \|x - \alpha\|v\|^{-1}v + \alpha(\|v\|^{-1}v - \xi) - \bar{x}\|^2 \\
\|x - \alpha\|v\|^{-1}v - \bar{x}\|^2 &+ \alpha^2\|\|v\|^{-1}v - \xi\|^2 \\
&+ 2\alpha\langle \|v\|^{-1}v - \xi, x - \alpha\|v\|^{-1}v - \bar{x} \rangle \\
&\le \|x - \alpha\|v\|^{-1}v - \bar{x}\|^2 \\
&+ \alpha^2\delta_f^2 + 2\alpha\delta_f(K_0 + \bar{K} + 1) \\
&\le \|x - \bar{x}\|^2 - 2\langle x - \bar{x}, \alpha\|v\|^{-1}v \rangle \\
&+ \alpha^2(1 + \delta_f)^2 + 2\alpha\delta_f(K_0 + \bar{K} + 1) \\
&< \|x - \bar{x}\|^2 - 2\alpha(4^{-1}\bar{L}^{-1}\epsilon) \\
&+ \alpha^2(1 + \delta_f^2) + 2\alpha\delta_f(K_0 + \bar{K} + 1) \\
&\le \|x - \bar{x}\|^2 - \alpha(4\bar{L})^{-1}\epsilon + 2\alpha^2. \qquad (2.149)
\end{aligned}
$$

In view of (2.7), (2.8), (2.15), (2.132), and (2.149),

$$\|y_0 - \bar{x}\|^2 \le (K_0 + \bar{K})^2 + 2$$

and

$$\|y_0 - \bar{x}\| \le K_0 + \bar{K} + 2. \qquad (2.150)$$

By (2.10), (2.131), (2.132), (2.140), and (2.148)–(2.150),

$$\|y - \bar{x}\|^2 = \|y - P_C(x - \alpha\xi) + P_C(x - \alpha\xi) - \bar{x}\|^2$$

$$\|y - P_C(x - \alpha\xi)\|^2 + \|P_C(x - \alpha\xi) - \bar{x}\|^2 + 2\|y - P_C(x - \alpha\xi)\|\|P_C(x - \alpha\xi) - \bar{x}\|$$
$$\leq \|y_0 - \bar{x}\|^2 + \delta_C^2 + 2\delta_C\|y_0 - \bar{x}\|$$
$$\leq \|x - \bar{x}\|^2 - \alpha(4\bar{L})^{-1}\epsilon$$
$$+2\alpha^2 + \delta_C^2 + 2\delta_C(K_0 + \bar{K} + 2).$$

This completes the proof of Lemma 2.24.

Applying Lemma 2.24 with $\epsilon = 16\bar{L}$, we obtain the following result.

Lemma 2.25 *Let*

$$\bar{x} \in C_{min},$$

$K_0 > 0, \alpha \in (0, 1], \delta_f, \delta_C \in (0, 1]$ *satisfy*

$$\delta_f(K_0 + \bar{K} + 1) \leq 2$$

; let a point $x \in X$ satisfy $\|x\| \leq K_0$, $f(x) > \inf(f, C) + 16\bar{L}$, and $v \in \partial_{4\bar{L}} f(x)$. Then

$$\|v\| \geq 12\bar{L}(K_0 + \bar{K})^{-1}$$

and for each

$$\xi \in B_X(\|v\|^{-1}v, \delta_f),$$

and each

$$y \in B_X(P_C(x - \alpha\xi), \delta_C)$$

the following inequality holds:

$$\|y - \bar{x}\|^2 \leq \|x - \bar{x}\|^2 - 2\alpha + 2\delta_C(K_0 + \bar{K} + 3).$$

Lemma 2.24 implies the following result.

Lemma 2.26 *Let $K_0 > 0, \epsilon \in (0, 16\bar{L}], \alpha \in (0, 1], \delta_f, \delta_C \in (0, 1]$ satisfy*

$$\delta_f(K_0 + \bar{K} + 1) \leq (8\bar{L})^{-1}\epsilon,$$

let a point $x \in X$ satisfy $\|x\| \leq K_0$, $f(x) > \inf(f, C) + \epsilon$ and let $v \in \partial_{\epsilon/4} f(x)$. Then

$$\|v\| \geq 3(4K_0 + 4\bar{K})^{-1}\epsilon$$

and for each

$$\xi \in B_X(\|v\|^{-1}v, \delta_f),$$

and each

$$y \in B_X(P_C(x - \alpha\xi), \delta_C)$$

the following inequality holds:

$$d(y, C_{min})^2 \le d(x, C_{min})^2 - (4\bar{L})^{-1}\alpha\epsilon + 2\alpha^2 + \delta_C^2 + 2\delta_C(K_0 + \bar{K} + 2).$$

Lemma 2.27 *Let*

$$\bar{x} \in C_{min}, \tag{2.151}$$

$K_0 \ge \bar{K}, L_0 > 0,$

$$|f(z_1) - f(z_2)| \le L_0\|z_1 - z_2\| \text{ for all } z_1, z_2 \in B_X(0, K_0 + 1), \tag{2.152}$$

$$\epsilon \in (0, 16(L_0 + \bar{L})], \tag{2.153}$$

$\alpha \in (0, 1], \delta_f, \delta_C \in (0, 1]$ *satisfy*

$$\delta_f(K_0 + \bar{K} + 5L_0 + 5\bar{L} + 1) \le \epsilon/8, \tag{2.154}$$

let a point $x \in X$ satisfy

$$\|x\| \le K_0, \tag{2.155}$$

$$v \in \partial_{\epsilon/4}f(x), \tag{2.156}$$

$$\xi \in B_X(v, \delta_f), \tag{2.157}$$

and let

$$y \in B_X(P_C(x - \alpha\xi), \delta_C). \tag{2.158}$$

Then

$$\|y - \bar{x}\|^2 \le \|x - \bar{x}\|^2 + 2\alpha(f(\bar{x}) - f(x)) + 3\alpha\epsilon/4$$
$$+ 2\delta_C(K_0 + \bar{K} + 5L_0 + 5\bar{L} + 1) + 25(L_0 + \bar{L})^2\alpha^2.$$

Proof Set

$$y_0 = x - \alpha \xi. \tag{2.159}$$

Lemma 2.19, (2.152), (2.153), and (2.156) imply that

$$\|v\| \le 5L_0 + 4\bar{L}. \tag{2.160}$$

It follows from (2.8), (2.151), (2.155), (2.157), (2.159), and (2.160) that

$$
\begin{aligned}
\|y_0 - \bar{x}\|^2 &= \|x - \alpha\xi - \bar{x}\|^2 \\
&= \|x - \alpha v + (\alpha v - \alpha\xi) - \bar{x}\|^2 \\
&\le \|x - \alpha v - \bar{x}\|^2 + \alpha^2 \|v - \xi\|^2 + 2\alpha \langle v - \xi, x - \alpha v - \bar{x} \rangle \\
&\le \|x - \alpha v - \bar{x}\|^2 + \alpha^2 \delta_f^2 + 2\alpha \delta_f \|x - \alpha v - \bar{x}\| \\
&\le \|x - \alpha v - \bar{x}\|^2 + \alpha^2 \delta_f^2 + 2\alpha \delta_f (K_0 + \bar{K} + \alpha(5L_0 + 4\bar{L})) \\
&\le \|x - \bar{x}\|^2 - 2\alpha \langle x - \bar{x}, v \rangle + \alpha^2 (5L_0 + 4\bar{L})^2 \\
&\quad + \alpha^2 \delta_f^2 + 2\alpha \delta_f (K_0 + \bar{K} + \alpha(5L_0 + 4\bar{L})).
\end{aligned}
\tag{2.161}
$$

In view of (2.256),

$$\langle v, \bar{x} - x \rangle \le f(\bar{x}) - f(x) + \epsilon/4. \tag{2.162}$$

By (2.154), (2.161), and (2.62),

$$
\begin{aligned}
\|y_0 - \bar{x}\|^2 &\le \|x - \bar{x}\|^2 + 2\alpha(f(\bar{x}) - f(x)) + \alpha\epsilon/2 \\
&\quad + \alpha^2(5L_0 + 4\bar{L})^2 + \alpha^2 \delta_f^2 + 2\alpha \delta_f(K_0 + \bar{K} + \alpha(5L_0 + 4\bar{L})) \\
&\le \|x - \bar{x}\|^2 + 2\alpha(f(\bar{x}) - f(x)) + \alpha\epsilon/2 + 25\alpha^2(L_0 + \bar{L})^2 \\
&\quad + 2\alpha \delta_f(K_0 + \bar{K} + 5(L_0 + \bar{L})) \\
&\le \|x - \bar{x}\|^2 + 2\alpha(f(\bar{x}) - f(x)) + \alpha\epsilon/2 + 25\alpha^2(L_0 + \bar{L})^2 + \epsilon\alpha/4.
\end{aligned}
\tag{2.163}
$$

By (2.7), (2.8), (2.151), (2.153), and (2.163),

$$
\begin{aligned}
\|y_0 - \bar{x}\|^2 &\le \|x - \bar{x}\|^2 + \alpha\epsilon + 25(L_0 + \bar{L})^2\alpha^2 + 2\alpha L_0 \|\bar{x} - x\| \\
&\le (K_0 + \bar{K})^2 + 16(L_0 + \bar{L}) + (5(L_0 + \bar{L}))^2 \\
&\quad + 2L_0(K_0 + \bar{K}) \le (K_0 + \bar{K} + 5L_0 + 5\bar{L})^2,
\end{aligned}
$$

$$\|y_0 - \bar{x}\| \leq K_0 + \bar{K} + 5(L_0 + \bar{L}). \tag{2.164}$$

It follows from (2.10), (2.131), (2.151), (2.158), (2.159), and (2.163) that

$$
\begin{aligned}
\|y - \bar{x}\|^2 &= \|y - P_C(x - \alpha\xi) + P_C(x - \alpha\xi) - \bar{x}\|^2 \\
&\leq \|y - P_C(x - \alpha\xi)\|^2 + 2\|y - P_C(x - \alpha\xi)\|\|P_C(x - \alpha\xi) - \bar{x}\| \\
&\quad + \|P_C(x - \alpha\xi) - \bar{x}\|^2 \\
&\leq \delta_C^2 + 2\delta_C(K_0 + \bar{K} + 5L_0 + 5\bar{L}) + \|y_0 - \bar{x}\|^2 \\
&\leq \delta_C^2 + 2\delta_C(K_0 + \bar{K} + 5L_0 + 5\bar{L}) + \|x - \bar{x}\|^2 \\
&\quad + 2\alpha(f(\bar{x}) - f(x)) + 3\alpha\epsilon/4 + 25\alpha^2(L_0 + \bar{L})^2 \\
&\leq \|x - \bar{x}\|^2 + 2\alpha(f(\bar{x}) - f(x)) + 3\alpha\epsilon/4 \\
&\quad + 25\alpha^2(L_0 + \bar{L})^2 + 2\delta_C(K_0 + \bar{K} + 5L_0 + 5\bar{L} + 1).
\end{aligned}
$$

Lemma 2.27 is proved.

2.6   Proof of Theorem 2.1

By (2.17), (2.19), and (2.21),

$$B_X(x_i, \delta_C) \cap C \neq \emptyset, \ i = 1, \dots, n. \tag{2.165}$$

In view if (2.12), (2.19), and (2.165),

$$\|x_i\| \leq K_1 + 1, \ i = 0, 1, \dots, n. \tag{2.166}$$

Fix

$$\bar{x} \in C_{min}. \tag{2.167}$$

It follows from (2.6), (2.7), and (2.167) that

$$\|\bar{x}\| \leq \bar{K} \leq K_1. \tag{2.168}$$

Assume that the assertion of the theorem does not hold. Then

$$f(x_i) > \inf(f, C) + \epsilon, \ i = 1, \dots, n. \tag{2.169}$$

Let $i \in \{1, \dots, n - 1\}$. In view of (2.12)–(2.15), (2.18), (2.20), (2.21), and (2.167), we apply Lemma 2.21 with

$$K_0 = K_1 + 1, \ L_0 = L_1, \ \xi = \xi_i, \ x = x_i, \ y = x_{i+1}$$

and obtain that

$$
\begin{aligned}
\|x_{i+1} - \bar{x}\|^2 &\le \|x_i - \bar{x}\|^2 - 2^{-1}\alpha\epsilon + \delta_C^2 \\
&\quad + 2\delta_C(K_0 + \bar{K} + 5L_1 + 5\bar{L} + 1) + 25(L_1 + \bar{L})^2\alpha^2 \\
&= \|x_i - \bar{x}\|^2 - 4^{-1}\alpha\epsilon + 2\delta_C(K_1 + \bar{K} + 5L_1 + 5\bar{L} + 2) \\
&\le \|x_i - \bar{x}\|^2 - 8^{-1}\alpha\epsilon.
\end{aligned}
\tag{2.170}
$$

Relations (2.166) and (2.168) imply that

$$(1 + K_1 + \bar{K})^2 \ge \|x_1 - \bar{x}\|^2 \ge \|x_1 - \bar{x}\|^2 - \|x_n - \bar{x}\|^2$$

$$= \sum_{i=1}^{n-1}(\|x_i - \bar{x}\|^2 - \|x_{i+1} - \bar{x}\|^2) \ge 8^{-1}\alpha\epsilon(n-1).$$

Together with (2.18), this implies that

$$
\begin{aligned}
n &\le 1 + 8(1 + K_1 + \bar{K})^2\alpha^{-1}\epsilon^{-1} \\
&= 1 + 800(1 + K_1 + \bar{K})^2(L_1 + \bar{L})^2\epsilon^{-2}.
\end{aligned}
$$

This contradicts (2.16). The contradiction we have reached completes the proof of Theorem 2.1.

2.7 Proof of Theorem 2.2

By (2.27), (2.30), and (2.32),

$$B_X(x_i, \delta_C) \cap C \ne \emptyset, \ i = 1, \dots, n. \tag{2.171}$$

Fix

$$\bar{x} \in C_{min}. \tag{2.172}$$

It follows from (2.7), (2.8), and (2.172) that

$$\|\bar{x}\| \le \bar{K}. \tag{2.173}$$

It follows from (2.10), (2.29), (2.30) and (2.172) that

$$\|x_1 - \bar{x}\| \le \|x_1 - P_0(x_0)\| + \|P_0(x_0) - \bar{x}\|$$

$$\le \delta_C + \|x_0 - \bar{x}\| \le 1 + K_1 + \bar{K}. \tag{2.174}$$

In view of (2.172)–(2.174),

$$\|x_1\| \le 1 + K_1 + 2\bar{K} \le 1 + 3K_1, \tag{2.175}$$

$$d(x_1, C_{min}) \le 1 + K_1 + \bar{K}. \tag{2.176}$$

Assume that the assertion of the theorem does not hold. Then

$$f(x_i) > \inf(f, C) + \epsilon, \ i = 1, \ldots, n. \tag{2.177}$$

We show by induction that for $i = 1, \ldots, n$,

$$d(x_i, C_{min}) \le 1 + K_1 + \bar{K}. \tag{2.178}$$

Assume that an integer i satisfies $1 \le i < n$ and that (2.178) is true. (In view of (2.176), our assumption holds with $i = 1$.) By (2.7), (2.8), and (2.178),

$$\|x_i\| \le 1 + K_1 + 2\bar{K} \le 1 + 3K_1. \tag{2.179}$$

There are two cases:

$$f(x_i) \le \inf(f, C) + 4; \tag{2.180}$$

$$f(x_i) > \inf(f, C) + 4. \tag{2.181}$$

Assume that (2.180) holds. In view of (2.7), (2.8), and (2.180),

$$\|x_i\| \le \bar{K}. \tag{2.182}$$

It follows from (2.10), (2.32), (2.172), (2.173), and (2.182) that

$$\|x_{i+1} - \bar{x}\| \le \|x_{i+1} - P_i(x_i - \alpha\xi_i)\| + \|P_i(x_i - \alpha\xi_i) - \bar{x}\|$$
$$\le \delta_C + \|x_i - \alpha\xi_i - \bar{x}\| \le 1 + \|x_i - \bar{x}\| + \alpha\|\xi_i\|$$
$$\le 1 + 2\bar{K} + \alpha\|\xi_i\|. \tag{2.183}$$

Lemma 2.19, (2.8), (2.9), and (2.182) imply that

$$\partial_{\epsilon/4} f(x_i) \subset B_X(0, \bar{L} + \epsilon/4). \tag{2.184}$$

By (2.23)–(2.25), (2.31), (2.98), and (2.184),

$$\alpha \| \xi_i \| \leq \alpha(\bar{L} + \epsilon/4 + \delta_f) \leq \alpha(\bar{L} + 2) \leq 1. \tag{2.185}$$

In view of (2.183) and (2.185),

$$\| x_{i+1} - \bar{x} \| \leq 2 + 2\bar{K} \leq 1 + K_1 + \bar{K}$$

and

$$d(x_{i+1}, C_{min}) \leq 1 + K_1 + \bar{K}.$$

Assume that (2.181) holds. In view of (2.23)–(2.25), (2.31), (2.32), (2.48), (2.178), (2.179), and (2.181), we apply Lemma 2.23 with

$$P_C = P_i, \ K_0 = 3K_1 + 1, \ L_0 = L_1, \ \xi = \xi_i, \ x = x_i, \ y = x_{i+1}$$

and obtain that

$$\begin{aligned}
d(x_{i+1}, C_{min})^2 &\leq d(x_i, C_{min})^2 - 2\alpha \\
&\quad + 2\delta_C(3K_1 + \bar{K} + 2 + 5L_1 + 5\bar{L}) + 25(L_1 + \bar{L})^2\alpha^2 \\
&\leq d(x_i, C_{min})^2 - \alpha + 2\delta_C(3K_1 + \bar{K} + 2 + 5L_1 + 5\bar{L}) \\
&\leq d(x_i, C_{min})^2 - \alpha/2 \leq d(x_i, C_{min})^2
\end{aligned}$$

and

$$d(x_{i+1}, C_{min}) \leq d(x_i, C_{min}) \leq 1 + K_1 + \bar{K}.$$

Thus in the both cases

$$d(x_{i+1}, C_{min}) \leq K_1 + \bar{K} + 1.$$

Thus we showed by induction that for all integers $i = 1, \ldots, n$, (2.178) holds and that

$$\| x_i \| \leq 1 + K_1 + 2\bar{K} \leq 1 + 3K_1. \tag{2.186}$$

In view of (2.23)–(2.25), (2.28), (2.31), (2.32), (2.171), (2.172), (2.177), and (2.186), we apply Lemma 2.21 with

$$P_C = P_i, \ K_0 = 3K_1 + 1, \ L_0 = L_1, \ \xi = \xi_i, \ x = x_i, \ y = x_{i+1}$$

and obtain that

$$\|x_{i+1} - \bar{x}\|^2 \le \|x_i - \bar{x}\|^2 - 2^{-1}\alpha\epsilon$$

$$+2\delta_C(3K_1 + \bar{K} + 2 + 5L_1 + 5\bar{L}) + 25(L_1 + \bar{L})^2\alpha^2$$

$$\le \|x_i - \bar{x}\|^2 - 4^{-1}\alpha\epsilon + 2\delta_C(3K_1 + \bar{K} + 2 + 5L_1 + 5\bar{L})$$

$$\le \|x_i - \bar{x}\|^2 - 8^{-1}\alpha\epsilon. \tag{2.187}$$

By (2.28), (2.175), and (2.187),

$$(1 + K_1 + \bar{K})^2 \ge \|x_1 - \bar{x}\|^2$$

$$\ge \|x_1 - \bar{x}\|^2 - \|x_n - \bar{x}\|^2$$

$$= \sum_{i=1}^{n-1}(\|x_i - \bar{x}\|^2 - \|x_{i+1} - \bar{x}\|^2) \ge 8^{-1}\alpha\epsilon(n - 1)$$

and

$$n \le 1 + 8(1 + K_1 + \bar{K})^2(\epsilon\alpha)^{-1}$$

$$\le 1 + 800(1 + K_1 + \bar{K})^2\epsilon^{-2}(L_1 + \bar{L})^2.$$

This contradicts (2.26). The contradiction we have reached completes the proof of Theorem 2.2.

2.8   Proof of Theorem 2.3

Set

$$P_n = P_{n-1}, \ \xi_n \in \partial f(x_n), \ \alpha_n = 1, \tag{2.188}$$

$$x_{n+1} = P_n(x_n - \xi_n). \tag{2.189}$$

By (2.37), (2.39), (2.41), (2.188), and (2.189),

$$B_X(x_i, \delta_C) \cap C \ne \emptyset, \ i = 1, \ldots, n + 1. \tag{2.190}$$

In view if (2.33), (2.38), and (2.190),

$$\|x_i\| \le K_1 + 1, \ i = 0, 1, \ldots, n + 1. \tag{2.191}$$

Fix

$$\bar{x} \in C_{min}. \tag{2.192}$$

It follows from (2.7), (2.8), and (2.192) that

$$\|\bar{x}\| \leq \bar{K} \leq K_1. \tag{2.193}$$

Let $i \in \{1, \ldots, n\}$. In view of (2.34), (2.36), (2.40), (2.41), (2.188), (2.189), (2.191), and (2.192), we apply Lemma 2.27 with

$$K_0 = K_1 + 1, \ L_0 = L_1, \ \alpha = \alpha_i, \ \xi = \xi_i, \ x = x_i, \ y = x_{i+1}$$

and obtain that

$$\|x_{i+1} - \bar{x}\|^2 \leq \|x_i - \bar{x}\|^2 + 2\alpha_i(f(\bar{x}) - f(x_i)) + 3\alpha_i\epsilon/4$$
$$+25(L_1 + \bar{L})^2\alpha_i^2 + 2\delta_C(K_1 + \bar{K} + 5L_1 + 5\bar{L} + 2). \tag{2.194}$$

Inequality (2.194) implies that

$$\alpha_i(f(x_i) - f(\bar{x})) \leq 2^{-1}\|x_i - \bar{x}\|^2 - 2^{-1}\|x_{i+1} - \bar{x}\|^2 + 3\alpha_i\epsilon/4$$
$$+25(L_1 + \bar{L})^2\alpha_i^2 + 2\delta_C(K_1 + \bar{K} + 5L_1 + 5\bar{L} + 2). \tag{2.195}$$

In view of (2.195),

$$\sum_{i=1}^{n}\alpha_i(f(x_i) - f(\bar{x})) \leq 2^{-1}\|x_1 - \bar{x}\|^2 - 2^{-1}\|x_{n+1} - \bar{x}\|^2$$

$$+3\left(\sum_{i=1}^{n}\alpha_i\right)\epsilon/4 + 25(L_1 + \bar{L})^2\sum_{i=1}^{n}\alpha_i^2 + 2n\delta_C(K_1 + \bar{K} + 5L_1 + 5\bar{L} + 2).$$

$$\tag{2.196}$$

It follows from (2.191), (2.193), and (2.196) the convexity of f that

$$\min\{f(x_i) : t = 1, \ldots, n\} - \inf(f, C), \ f\left(\left(\sum_{j=1}^{n}\alpha_j\right)^{-1}\sum_{i=1}^{n}\alpha_i x_i\right) - \inf(f, C)$$

$$\leq \left(\sum_{j=1}^{n}\alpha_j\right)^{-1}\sum_{i=1}^{n}\alpha_i(f(x_i) - \inf(f, C))$$

$$\leq \left(\sum_{i=1}^{n}\alpha_i\right)^{-1}(2K_1 + 1)^2 + 3\epsilon/4 + 25(L_1 + \bar{L})^2\left(\sum_{i=1}^{n}\alpha_i^2\right)\left(\sum_{i=1}^{n}\alpha_i\right)^{-1}$$

$$+2n \left(\sum_{i=1}^{n} \alpha_i \right)^{-1} \delta_C (\bar{K} + K_1 + 5L_1 + 5\bar{L} + 2).$$

Theorem 2.3 is proved.

2.9 Proof of Theorem 2.5

Set

$$P_n = P_{n-1}, \ \xi_n \in \partial f(x_n), \tag{2.197}$$

$$x_{n+1} = P_n(x_n - \alpha \xi_n). \tag{2.198}$$

Fix

$$\bar{x} \in C_{min}. \tag{2.199}$$

It follows from (2.7), (2.8), and (2.199) that

$$\|\bar{x}\| \leq \bar{K}. \tag{2.200}$$

It follows from (2.10), (2.47), (2.48), and (2.200) that

$$\|x_1 - \bar{x}\| \leq \|x_1 - P_0(x_0)\| + \|P_0(x_0) - \bar{x}\|$$
$$\leq \delta_C + \|x_0 - \bar{x}\| \leq 1 + K_1 + \bar{K}. \tag{2.201}$$

We show that for all $i = 1, \ldots, n+1$,

$$\|x_i - \bar{x}\| \leq 1 + K_1 + \bar{K}. \tag{2.202}$$

Assume that an integer i satisfies $1 \leq i < n+1$ and that (2.202) is true. (In view of (2.201), our assumption holds with $i = 1$.) There are two cases:

$$f(x_i) \leq \inf(f, C) + 4; \tag{2.203}$$

$$f(x_i) > \inf(f, C) + 4. \tag{2.204}$$

Assume that (2.203) holds. In view of (2.7), (2.8), and (2.203),

$$\|x_i\| \leq \bar{K}. \tag{2.205}$$

Lemma 2.19, (2.7)–(2.9), (2.42), and (2.205) imply that

$$\partial_{\epsilon/4} f(x_i) \subset B_X(0, \bar{L} + \epsilon/4) \subset B_X(0, \bar{L} + 1). \tag{2.206}$$

By (2.49), (2.197), and (2.206),

$$\|\xi_i\| \leq \bar{L} + 2. \tag{2.207}$$

It follows from (2.10), (2.43), (2.197), (2.198), (2.200), (2.205), and (2.207) that

$$\begin{aligned}
\|x_{i+1} - \bar{x}\| &\leq \|x_{i+1} - P_i(x_i - \alpha\xi_i)\| + \|P_i(x_i - \alpha\xi_i) - \bar{x}\| \\
&\leq \delta_C + \|x_i - \alpha\xi_i - \bar{x}\| \leq 1 + \|x_i - \bar{x}\| + \alpha\|\xi_i\| \\
&\leq 1 + 2\bar{K} + \alpha(\bar{L} + 2) \leq 1 + 2\bar{K} + 2 \leq 1 + \bar{K} + K_1.
\end{aligned}$$

Assume that (2.204) holds. In view of (2.42)–(2.45), (2.50), (2.147), (2.148), (2.199), (2.202), and (2.204), we apply Lemma 2.27 with

$$K_0 = 1 + 2\bar{K} + K_1, \ L_0 = L_1, \ \xi = \xi_i, \ x = x_i, \ y = x_{i+1}, \epsilon = 4$$

and obtain that

$$\begin{aligned}
\|x_{i+1} - \bar{x}\|^2 &\leq \|x_i - \bar{x}\|^2 + 2\alpha(f(\bar{x}) - f(x_i)) + 3\alpha\epsilon/4 \\
&\quad + 25(L_1 + \bar{L})^2\alpha^2 + 2\delta_C(3\bar{K} + K_1 + 2 + 5L_1 + 5\bar{L}) \\
&\leq \|x_i - \bar{x}\|^2 - 8\alpha + 3\alpha + 25(L_1 + \bar{L})^2\alpha^2 + 2\delta_C(3\bar{K} + K_1 + 2 + 5L_1 + 5\bar{L}) \\
&\leq \|x_i - \bar{x}\|^2 - 2\alpha
\end{aligned}$$

and

$$\|x_{i+1} - \bar{x}\| \leq \|x_i - \bar{x}\| \leq 1 + \bar{K} + K_1. \tag{2.208}$$

Thus in both cases, (2.208) holds. Therefore by induction we showed that

$$\|x_i - \bar{x}\| \leq 1 + K_1 + \bar{K}, \ i = 0, \ldots, n + 1, \tag{2.209}$$

$$\|x_i\| \leq 1 + K_1 + 2\bar{K}, \ i = 0, \ldots, n + 1. \tag{2.210}$$

Let $i \in \{1, \ldots, n\}$. In view of (2.42), (2.45), (2.49), (2.50), (2.147), (2.148), and (2.210), we apply Lemma 2.27 with

$$K_0 = 2\bar{K} + K_1 + 1, \ L_0 = L_1, \ \xi = \xi_i, \ x = x_i, \ y = x_{i+1}$$

and obtain that

$$\|x_{i+1} - \bar{x}\|^2 \le \|x_i - \bar{x}\|^2 + 2\alpha(f(\bar{x}) - f(x_i)) + 3\alpha\epsilon/4$$

$$+2\delta_C(K_1 + 3\bar{K} + 5L_1 + 5\bar{L} + 2) + 25(L_1 + \bar{L})^2\alpha^2.$$

This inequality implies that

$$2\alpha(f(x_i) - f(\bar{x})) \le \|x_i - \bar{x}\|^2 - \|x_{i+1} - \bar{x}\|^2 + 3\alpha\epsilon/4$$

$$+25(L_1 + \bar{L})^2\alpha^2 + 2\delta_C(K_1 + 3\bar{K} + 5L_1 + 5\bar{L} + 2). \tag{2.211}$$

By (2.211),

$$2\alpha \sum_{i=1}^{n}(f(x_i) - f(\bar{x})) \le \sum_{i=1}^{n}(\|x_i - \bar{x}\|^2 - \|x_{i+1} - \bar{x}\|^2)$$

$$+n[3\alpha\epsilon/4 + 25(L_1 + \bar{L})^2\alpha^2 + 2\delta_C(K_1 + 3\bar{K} + 5L_1 + 5\bar{L} + 2)]$$

$$\le \|x_1 - \bar{x}\|^2 + n[3\alpha\epsilon/4 + 25(L_1 + \bar{L})^2\alpha^2 + 2\delta_C(K_1 + 3\bar{K} + 5L_1 + 5\bar{L} + 2)].$$

$$\tag{2.212}$$

It follows from (2.209) and (2.212) that

$$n^{-1}\sum_{i=1}^{n-1} f(x_i) - \inf(f, C) \le (2\alpha)^{-1}(1 + K_1 + \bar{K})^2 + 3\epsilon/8$$

$$+2^{-1}25(L_1 + \bar{L})^2\alpha + \alpha^{-1}\delta_C(K_1 + 3\bar{K} + 5L_1 + 5\bar{L} + 2).$$

Theorem 2.5 is proved.

2.10   Proof of Theorem 2.8

Fix

$$\bar{x} \in C_{min}. \tag{2.213}$$

By (2.51), (2.56), (2.58), (2.59), and (2.61), for every $i \in \{1, \ldots, n\}$,

$$\|x_i\| \le K_1 + 1. \tag{2.214}$$

Assume that an integer $i \in \{1, \ldots, n\} \setminus \{n\}$ satisfies

$$f(x_i) > \inf(f, C) + \epsilon. \tag{2.215}$$

We apply Lemma 2.14 with

$$P_C = P_i, \ K_0 = K_1, \ \xi = \xi_i, \ x = x_i, \ y = x_{i+1}, \ \alpha = \alpha_i$$

and in view of (2.52), (2.53), (2.54), (2.57), (2.58), (2.60), (2.61), and (2.215), obtain that

$$\partial_{\epsilon/4} f(x_i) \subset \{v \in X : \ \|v\| \geq (3/4)\epsilon(K_1 + \bar{K})^{-1}\}, \tag{2.216}$$

$$\|\xi_i\| \geq (3/4)\epsilon(K_1 + \bar{K})^{-1} - \delta_f,$$

$$\begin{aligned}
\|x_{i+1} - \bar{x}\|^2 &\leq \|x_i - \bar{x}\|^2 - (4\bar{L})^{-1}\alpha_i\epsilon + 2\alpha_i^2 + \delta_C^2 + 2\delta_C(K_1 + \bar{K} + 2) \\
&\leq \|x_i - \bar{x}\|^2 - (8\bar{L})^{-1}\alpha_i\epsilon + 2\delta_C(K_1 + \bar{K} + 3) \\
&\leq \|x_i - \bar{x}\|^2 - (16\bar{L})^{-1}\alpha_i\epsilon.
\end{aligned}$$

Thus we have shown that the following property holds:

(a) if an integer $i \in \{1, \ldots, n\} \setminus \{n\}$ and

$$f(x_i) > \inf(f, C) + \epsilon$$

then (2.216) holds and

$$\|x_{i+1} - \bar{x}\|^2 \leq \|x_i - \bar{x}\|^2 - (16\bar{L})^{-1}\alpha_i\epsilon. \tag{2.217}$$

Assume that an integer $j \in \{1, \ldots, n\} \setminus \{n\}$ and that

$$f(x_i) > \inf(f, C) + \epsilon, \ i = 1, \ldots, j. \tag{2.218}$$

Property (a) implies that for all $i = 1, \ldots, j$ (2.216) and (2.217) hold. Combined with (2.7), (2.8), (2.57), (2.213), and (2.214), this implies that

$$\begin{aligned}
(K_1 + \bar{K} + 1)^2 &\geq \|x_1 - \bar{x}\|^2 \geq \|x_1 - \bar{x}\|^2 - \|x_{j+1} - \bar{x}\|^2 \\
&= \sum_{i=1}^{j}(\|x_i - \bar{x}\|^2 - \|x_{i+1} - \bar{x}\|^2) \\
&\geq (16\bar{L})^{-1}\epsilon(j+1)\min\{\alpha_i : \ i = 1, \ldots, j\} \geq 2^{-1}(16\bar{L})^{-2}\epsilon^2(j+1)
\end{aligned}$$

and

$$j + 1 \leq 2(1 + K_1 + \bar{K})^2\epsilon^{-2}(16\bar{L})^2 \leq n - 2.$$

This implies that there exists $j \in \{1, \ldots, n\}$ such that

$$f(x_j) \le \inf(f, C) + \epsilon$$

and if $i \in \{1, \ldots, n\} \setminus \{j\}$, then

$$f(x_i) > \inf(f, C) + \epsilon$$

and

$$\partial_{\epsilon/4} f(x_i) \subset \{v \in X : \|v\| \ge (3/4)\epsilon (K_1 + \bar{K})^{-1}\}.$$

Theorem 2.8 is proved.

2.11 Proof of Theorem 2.10

In view of (2.66), (2.69), and (2.71),

$$B_X(x_i, \delta_C) \cap C \ne \emptyset, \ i = 1, \ldots, n. \tag{2.219}$$

Fix

$$\bar{x} \in C_{min}. \tag{2.220}$$

It follows from (2.7), (2.8), and (2.220) that

$$\|\bar{x}\| \le \bar{K}. \tag{2.221}$$

It follows from (2.10), (2.68), (2.69), (2.220), and (2.221) that

$$\|x_1 - \bar{x}\| \le \|x_1 - P_0(x_0)\| + \|P_0(x_0) - \bar{x}\|$$
$$\le 1 + \|x_0 - \bar{x}\| \le 1 + K_1 + \bar{K}. \tag{2.222}$$

We show that for all $i = 1, \ldots, n$,

$$\|x_i - \bar{x}\| \le 1 + K_1 + \bar{K}. \tag{2.223}$$

Assume that an integer i satisfies $1 \le i < n$ and that (2.223) is true. (In view of (2.222), our assumption holds with $i = 1$.) We show that our assumption holds for $i + 1$ too. There are two cases:

$$f(x_i) \le \inf(f, C) + 4; \tag{2.224}$$

$$f(x_i) > \inf(f, C) + 4. \tag{2.225}$$

Assume that (2.224) holds. In view of (2.7), (2.8), and (2.224),

$$\|x_i\| \le \bar{K}. \tag{2.226}$$

It follows from (2.10), (2.62), (2.64), (2.67), (2.70), (2.71), (2.216), and (2.220) that

$$\|x_{i+1} - \bar{x}\| \le \|x_{i+1} - P_i(x_i - \alpha_i \xi_i)\| + \|P_i(x_i - \alpha_i \xi_i) - \bar{x}\|$$
$$\le \delta_C + \|x_i - \alpha_i \xi_i - \bar{x}\| \le \delta_C + \|x_i - \bar{x}\| + \|\xi_i\|$$
$$\le \delta_C + 1 + 2\bar{K} + \delta_f \le 3 + 2\bar{K} \le 1 + \bar{K} + K_1.$$

Assume that (2.225) holds. In view of (2.62)–(2.64), (2.67), (2.70), (2.71), (2.220), (2.221), (2.223), and (2.225), we apply Lemma 2.24 with

$$P_C = P_i, \;\; K_0 = 1 + 2\bar{K} + K_1, \;\; \alpha = \alpha_i, \;\; \xi = \xi_i, \;\; x = x_i, \;\; y = x_{i+1}, \epsilon = 4$$

and obtain that

$$\|x_{i+1} - \bar{x}\|^2 \le \|x_i - \bar{x}\|^2 - \alpha_i \bar{L}^{-1} + 2\alpha_i^2 + \delta_C^2 + 2\delta_C(3\bar{K} + K_1 + 3)$$
$$\le \|x_i - \bar{x}\|^2 - (2\bar{L})^{-1}\alpha_i + 2\delta_C(3\bar{K} + K_1 + 4)$$
$$\le \|x_i - \bar{x}\|^2 - (4\bar{L})^{-1}\alpha_i.$$

By the relation above,

$$\|x_{i+1} - \bar{x}\| \le \|x_i - \bar{x}\| \le 1 + \bar{K} + K_1.$$

Thus in the both cases

$$\|x_{i+1} - \bar{x}\| \le 1 + K_1 + \bar{K}.$$

Therefore by induction we showed that (2.223) holds for all $i = 1, \ldots, n$. This implies that for all $i = 1, \ldots, n$,

$$\|x_i\| \le 1 + K_1 + 2\bar{K}.$$

Assume that $i \in \{1, \ldots, n\} \setminus \{n\}$ and that

$$f(x_i) > \inf(f, C) + \epsilon.$$

In view of the inequality above, (2.62)–(2.64), (2.67), (2.70), (2.71), (2.220), and (2.223), we apply Lemma 2.24 with

$$P_C = P_i, \ K_0 = 2\bar{K} + K_1 + 1, \ \alpha = \alpha_i, \ \xi = \xi_i, \ x = x_i, \ y = x_{i+1}$$

and obtain that

$$\partial_{\epsilon/4} f(x_i) \subset \{v \in X : \ \|v\| \geq (3/4)\epsilon(K_1 + 3\bar{K} + 1)^{-1}\}, \tag{2.227}$$

$$\|x_{i+1} - \bar{x}\|^2 \leq \|x_i - \bar{x}\|^2 - (4\bar{L})^{-1}\alpha_i\epsilon + 2\alpha_i^2 + \delta_C^2 + 2\delta_C(K_1 + 3\bar{K} + 3)$$
$$\leq \|x_i - \bar{x}\|^2 - (8\bar{L})^{-1}\alpha_i\epsilon + 2\delta_C(K_1 + 3\bar{K} + 4)$$
$$\leq \|x_i - \bar{x}\|^2 - (16\bar{L})^{-1}\alpha_i\epsilon.$$

Thus we have shown that the following property holds:

(a) if an integer $i \in \{1, \ldots, n\} \setminus \{n\}$ and

$$f(x_i) > \inf(f, C) + \epsilon$$

then (2.227) holds and

$$\|x_{i+1} - \bar{x}\|^2 \leq \|x_i - \bar{x}\|^2 - (16\bar{L})^{-1}\alpha_i\epsilon.$$

Assume that an integer $j \in \{1, \ldots, n\} \setminus \{n\}$ and that

$$f(x_i) > \inf(f, C) + \epsilon, \ i = 1, \ldots, j.$$

Property (a) and the inequality above imply that for all $i = 1, \ldots, j$ (2.227) holds and

$$\|x_{i+1} - \bar{x}\|^2 \leq \|x_i - \bar{x}\|^2 - 2^{-1}(16\bar{L})^{-2}\epsilon^2 \tag{2.228}$$

is true. By (2.65), (2.222), and (2.228),

$$(K_1 + \bar{K} + 1)^2 \geq \|x_1 - \bar{x}\|^2 \geq \|x_1 - \bar{x}\|^2 - \|x_{j+1} - \bar{x}\|^2$$
$$= \sum_{i=1}^{j} (\|x_i - \bar{x}\|^2 - \|x_{i+1} - \bar{x}\|^2)$$
$$\geq 2^{-1}j(16\bar{L})^{-2}\epsilon^2$$

and

$$j \leq \lfloor 2(1 + K_1 + \bar{K})^2\epsilon^{-2}(16\bar{L})^2 \rfloor \leq n - 2.$$

This implies that there exists $j \in \{1, \ldots, n\}$ such that

$$f(x_j) \leq \inf(f, C) + \epsilon$$

and if $i \in \{1, \ldots, j\} \setminus \{j\}$, then

$$f(x_i) > \inf(f, C) + \epsilon\}$$

and

$$\partial_{\epsilon/4} f(x_i) \subset \{v \in X : \|v\| \geq (3/4)\epsilon(K_1 + 3\bar{K} + 1)^{-1}\}.$$

Theorem 2.10 is proved.

2.12 An Auxiliary Result for Theorems 2.11–2.15

Assume that all the assumptions made in Sections 2.1–2.3 hold.

Proposition 2.28 *Let $\epsilon \in (0, 1]$. Then for each $x \in X$ satisfying*

$$d(x, C) < \min\{2^{-1}\bar{L}^{-1}\phi(\epsilon/2), \ \epsilon/2\}, \tag{2.229}$$

$$f(x) \leq \inf(f, C) + \min\{2^{-1}\phi(\epsilon/2), \ \epsilon/2\}, \tag{2.230}$$

the inequality $d(x, C_{min}) \leq \epsilon$ holds.

Proof In view of the definition of ϕ, $\phi(\epsilon/2) \in (0, 1]$ and

$$\text{if } x \in C \text{ satisfies } f(x) < \inf(f, C) + \phi(\epsilon/2),$$

$$\text{then } d(x, C_{min}) \leq \min\{1, \epsilon/2\}. \tag{2.231}$$

Assume that a point $x \in X$ satisfies (2.229) and (2.230). There exists a point $y \in C$ which satisfies

$$\|x - y\| < 2^{-1}\bar{L}^{-1}\phi(\epsilon/2) \text{ and } \|x - y\| < \epsilon/2. \tag{2.232}$$

Relations (2.7), (2.8), (2.230), and (2.232) imply that

$$x \in B_X(0, \bar{K}), \ y \in B_X(0, \bar{K} + 1). \tag{2.233}$$

By (2.223), (2.232), and the definition of $\bar{L}$ (see (2.9)),

$$|f(x) - f(y)| \leq \bar{L}\|x - y\| < \phi(\epsilon/2)2^{-1}. \tag{2.234}$$

It follows from the choice of the point y, (2.230), and (2.234) that $y \in C$ and

$$f(y) < f(x) + \phi(\epsilon/2)2^{-1} \le \inf(f, C) + \phi(\epsilon/2).$$

Combined with (2.231), this implies that $d(y, C_{min}) \le \epsilon/2$. Together with (2.232), this implies that

$$d(x, C_{min}) \le \|x - y\| + d(y, C_{min}) \le \epsilon.$$

This completes the proof of Proposition 2.28.

2.13 Proof of Theorem 2.11

We may assume without loss of generality that $\epsilon < 1$. In view of Proposition 2.28, there exist a number

$$\bar{\epsilon} \in (0, \epsilon/8) \tag{2.235}$$

such that

$$\text{if } x \in X, \ d(x, C) \le 2\bar{\epsilon}, \text{and } f(x) \le \inf(f, C) + 2\bar{\epsilon},$$

$$\text{then } d(x, C_{min}) \le \epsilon. \tag{2.236}$$

Fix

$$\bar{x} \in C_{min} \tag{2.237}$$

and

$$\epsilon_0 \in (0, 4^{-1}\bar{\epsilon}). \tag{2.238}$$

Since $\lim_{i \to \infty} \alpha_i = 0$ (see (2.73)), there is an integer $p_0 > 0$ such that

$$\bar{K} + 4 < p_0 \tag{2.239}$$

and that for all integers $p \ge p_0 - 1$, we have

$$\alpha_p < (32\bar{L})^{-1}\epsilon_0. \tag{2.240}$$

Since $\sum_{i=0}^{\infty} \alpha_i = \infty$ (see (2.73)), there exist a natural number $n_0 > p_0 + 4$ such that

$$\sum_{i=p_0}^{n_0-1} \alpha_i > (4p_0 + M + \|\bar{x}\|)^2 \epsilon_0^{-1} 16\bar{L}. \tag{2.241}$$

Fix

$$K_* > \bar{K} + 4 + M + 4n_0 + 4\|\bar{x}\| \tag{2.242}$$

and a positive number δ such that

$$6\delta(K_* + 1) < (16\bar{L})^{-1}\epsilon_0. \tag{2.243}$$

Assume that an integer $n \geq n_0$ and that

$$\{P_k\}_{k=0}^{n-1} \subset \mathcal{M}, \ \{x_k\}_{k=0}^{n} \subset X,$$

$$\|x_0\| \leq M, \ P_i(X) = C, \ i = 0, \ldots, n-1, \tag{2.244}$$

$$v_k \in \partial_\delta f(x_k) \setminus \{0\}, \ k = 0, \ldots, n-1, \tag{2.245}$$

$$\{\eta_k\}_{k=0}^{n-1}, \ \{\xi_k\}_{k=0}^{n-1} \subset B_X(0, \delta), \tag{2.246}$$

and that for all integers $k = 0, \ldots, n-1$, we have

$$x_{k+1} = P_k(x_k - \alpha_k \|v_k\|^{-1} v_k - \alpha_k \xi_k) - \alpha_k \eta_k. \tag{2.247}$$

In order to prove the theorem, it is sufficient to show that

$$d(x_k, C_{min}) \leq \epsilon \text{ for all integers } k \text{ satisfying } n_0 \leq k \leq n.$$

Assume that an integer

$$k \in [p_0, n-1], \tag{2.248}$$

$$\|x_k\| \leq K_*, \tag{2.249}$$

$$f(x_k) > \inf(f, C) + \epsilon_0. \tag{2.250}$$

In view of (2.237), (2.243), (2.245), (2.246), (2.247), (2.249), and (2.250), the conditions of Lemma 2.24 hold with $P_C = P_k$, $K_0 = K_*$, $\epsilon = \epsilon_0$, $\delta_f = \delta$, $\delta_C = \delta\alpha_k$, $\alpha = \alpha_k$, $x = x_k$, $v = v_k$, $\xi = \xi_k + \|v_k\|^{-1}v_k$, and $y = x_{k+1}$, and combined with (2.240), (2.243), and (2.248), this Lemma implies that

$$\|x_{k+1} - \bar{x}\|^2 \leq \|x_k - \bar{x}\|^2 - \alpha_k(4\bar{L})^{-1}\epsilon_0$$

$$+ 2\alpha_k^2 + \alpha_k^2 \delta^2 + 2\delta\alpha_k(K_* + \bar{K} + 2)$$
$$\leq \|x_k - \bar{x}\|^2 - \alpha_k(8\bar{L})^{-1}\epsilon_0 + 2\delta\alpha_k(K_* + \bar{K} + 3)$$
$$\leq \|x_k - \bar{x}\|^2 - \alpha_k(16\bar{L})^{-1}\epsilon_0.$$

Thus we have shown that the following property holds:

(P1)   If an integer $k \in [p_0, n-1]$ and (2.249) and (2.250) are valid, then we have

$$\|x_{k+1} - \bar{x}\|^2 \leq \|x_k - \bar{x}\|^2 - (16\bar{L})^{-1}\alpha_k\epsilon_0.$$

We claim that there exists an integer $j \in \{p_0, \ldots, n_0\}$ such that

$$f(x_j) \leq \inf(f, C) + \epsilon_0.$$

Assume the contrary. Then

$$f(x_i) > \inf(f, C) + \epsilon_0, \quad i = p_0, \ldots, n_0. \tag{2.251}$$

It follows from (2.10), (2.243), and (2.245)–(2.247)) that for all integers $i = 0, \ldots, n-1$, we have

$$\|x_{i+1} - \bar{x}\| \leq 1 + \|P_i(x_i - \alpha_i\|v_i\|^{-1}v_i - \alpha_i\xi_i) - \bar{x}\|$$
$$\leq 1 + \|x_i - \alpha_i\|v_i\|^{-1}v_i - \alpha_i\xi_i - \bar{x}\|$$
$$\leq 1 + \|x_i - \bar{x}\| + 2 = \|x_i - \bar{x}\| + 3. \tag{2.252}$$

By (2.242), (2.244), and (2.252), for all integers $i = 0, \ldots, n_0$,

$$\|x_i\| \leq \|x_0 - \bar{x}\| + 3i + \|\bar{x}\| \leq M + 3i + 2\|\bar{x}\| \leq M + 3n_0 + 2\|\bar{x}\| < K_*. \tag{2.253}$$

Let

$$i \in \{p_0, \ldots, n_0 - 1\}. \tag{2.254}$$

It follows from (2.251), (2.253), (2.254), and property (P1) that

$$\|x_{i+1} - \bar{x}\|^2 \leq \|x_i - \bar{x}\|^2 - (16\bar{L})^{-1}\alpha_i\epsilon_0. \tag{2.255}$$

Relations (2.253) and (2.255) imply that

$$(M + 3p_0 + \|\bar{x}\|)^2 \geq \|x_{p_0} - \bar{x}\|^2 - \|x_{n_0} - \bar{x}\|^2$$
$$= \sum_{i=p_0}^{n_0-1}[\|x_i - \bar{x}\|^2 - \|x_{i+1} - \bar{x}\|^2] \geq (16\bar{L})^{-1}\epsilon_0 \sum_{i=p_0}^{n_0-1}\alpha_i$$

and

$$\sum_{i=p_0}^{n_0-1} \alpha_i \le 16\bar{L}\epsilon_0^{-1}(M + 3p_0 + \|\bar{x}\|)^2.$$

This contradicts (2.241). The contradiction we have reached proves that there exists an integer

$$j \in \{p_0, \dots, n_0\} \tag{2.256}$$

such that

$$f(x_j) \le \inf(f, C) + \epsilon_0. \tag{2.257}$$

By (2.243), (2.244), (2.246), and (2.247), we have

$$d(x_j, C) \le \alpha_{j-1}\delta < \bar{\epsilon}. \tag{2.258}$$

In view of (2.236), (2.238), (2.257), and (2.258),

$$d(x_j, C_{min}) \le \epsilon. \tag{2.259}$$

We claim that for all integers i satisfying $j \le i \le n$,

$$d(x_i, C_{min}) \le \epsilon.$$

Assume the contrary. Then there exists an integer $k \in [j, n]$ for which

$$d(x_k, C_{min}) > \epsilon. \tag{2.260}$$

By (2.256), (2.259), and (2.260), we have

$$k > j \ge p_0. \tag{2.261}$$

By (2.259) we may assume without loss of generality that

$$d(x_i, C_{min}) \le \epsilon \text{ for all integers } i \text{ satisfying } j \le i < k. \tag{2.262}$$

Thus in view of (2.261) and (2.262)

$$d(x_{k-1}, C_{min}) \le \epsilon. \tag{2.263}$$

There are two cases:

$$f(x_{k-1}) \leq \inf(f, C) + \epsilon_0; \tag{2.264}$$

$$f(x_{k-1}) > \inf(f, C) + \epsilon_0. \tag{2.265}$$

Assume that (2.264) is valid. It follows from (2.7), (2.8), and (2.264) that

$$x_{k-1} \in X_0 \subset B_X(0, \bar{K}). \tag{2.266}$$

By (2.244), (2.246), and (2.247), there exists a point $z \in C$ such that

$$\|x_{k-1} - z\| \leq \delta. \tag{2.267}$$

By (2.10), (2.246), (2.247), and (2.267),

$$\|x_k - z\| \leq \alpha_{k-1}\delta + \|z - P_{k-1}(x_{k-1} - \alpha_{k-1}\|v_{k-1}\|^{-1}v_{k-1} - \alpha_{k-1}\xi_{k-1})\|$$

$$\leq \delta + \|z - x_{k-1}\| + \alpha_{k-1} + \delta = 3\delta + \alpha_{k-1}. \tag{2.268}$$

It follows from (2.240), (2.243), (2.261), and (2.268) that

$$d(x_k, C) \leq 3\delta + \alpha_{k-1} < \epsilon_0. \tag{2.269}$$

In view of (2.267) and (2.268),

$$\|x_k - x_{k-1}\| \leq \|x_k - z\| + \|z - x_{k-1}\| \leq 4\delta + \alpha_{k-1}. \tag{2.270}$$

It follows from (2.240), (2.243), (2.261), (2.263), and (2.270) that

$$d(x_k, C_{min}) \leq 2\epsilon. \tag{2.271}$$

Relations (2.7), (2.8), (2.263), and (2.271) imply that

$$x_{k-1}, x_k \in B_X(0, \bar{K} + 2).$$

Together with (2.9) and (2.270), the inclusion above implies that

$$|f(x_{k-1}) - f(x_k)| \leq \bar{L}\|x_{k-1} - x_k\| \leq \bar{L}(4\delta + \alpha_{k-1}). \tag{2.272}$$

In view of (2.240), (2.243), (2.264), and (2.272), we have

$$f(x_k) \leq f(x_{k-1}) + \bar{L}(4\delta + \alpha_{k-1})$$
$$\leq \inf(f, C) + \epsilon_0 + \bar{L}(4\delta + \alpha_{k-1}) \leq \inf(f, C) + 2\epsilon_0. \tag{2.273}$$

It follows from (2.236), (2.238), (2.269), and (2.273) that

$$d(x_k, C_{min}) \leq \epsilon.$$

This inequality contradicts (2.260). The contradiction we have reached proves (2.265).

By (2.7), (2.8), and (2.263), we have

$$\|x_{k-1}\| \leq \bar{K} + 1. \tag{2.274}$$

It follows from (2.240), (2.242), (2.243), (2.245), (2.247), (2.263), (2.265), and (2.274) that Lemma 2.26 holds with

$$x = x_{k-1}, \ y = x_k, \ \xi = \xi_{k-1}, \ v = v_{k-1}, \ \alpha = \alpha_{k-1}, \ K_0 = \bar{K} + 1,$$

$$\epsilon = \epsilon_0, \ \delta_f = \delta, \ \delta_C = \alpha_{k-1}\delta$$

and this implies that

$$\begin{aligned}
d(x_k, C_{min})^2 & \\
& \leq d(x_{k-1}, C_{min})^2 - \alpha_{k-1}(4\bar{L})^{-1}\epsilon_0 + 2\alpha_{k-1}^2 + \alpha_{k-1}^2\delta^2 + 2\alpha_{k-1}\delta(2\bar{K} + 3) \\
& \leq d(x_{k-1}, C_{min})^2 - (8\bar{L})^{-1}\alpha_{k-1}\epsilon_0 + 2\alpha_{k-1}\delta(2\bar{K} + 4) \\
& \leq d(x_{k-1}, C_{min})^2 - (16\bar{L})^{-1}\alpha_{k-1}\epsilon_0 \leq d(x_{k-1}, C_{min})^2 \leq \epsilon^2.
\end{aligned}$$

This contradicts (2.260).

The contradiction we have reached proves that $d(x_i, C_{min}) \leq \epsilon$ for all integers i satisfying $j \leq i \leq n$. Since $j \leq n_0$, this completes the proof of Theorem 2.11.

2.14 Proof of Theorem 2.12

We may assume that without loss of generality

$$\epsilon < 1, \ M > \bar{K} + 4. \tag{2.275}$$

Proposition 2.28 implies that there exists

$$\bar{\epsilon} \in (0, \epsilon/8) \tag{2.276}$$

such that

$$\text{if } x \in X, \ d(x, C) \leq 2\bar{\epsilon}, \ \text{and} f(x) \leq \inf(f, C) + 2\bar{\epsilon},$$

$$\text{then } d(x, C_{min}) \le \epsilon/4. \tag{2.277}$$

Set

$$\beta_0 = (64\bar{L})^{-1}\bar{\epsilon}. \tag{2.278}$$

Let

$$\beta_1 \in (0, \beta_0). \tag{2.279}$$

There exists an integer $n_0 \ge 4$ such that

$$\beta_1 n_0 > 16^2(3 + 2M)^2 \bar{\epsilon}^{-1} \bar{L}. \tag{2.280}$$

Fix

$$K_* > 2M + 4 + 4n_0 + 2\bar{K} + 2M \tag{2.281}$$

and a positive number δ such that

$$6\delta K_* < (64\bar{L})^{-1}\bar{\epsilon}\beta_1. \tag{2.282}$$

Fix a point

$$\bar{x} \in C_{min}. \tag{2.283}$$

Assume that an integer $n \ge n_0$,

$$\{P_i\}_{i=0}^{n-1} \subset \mathcal{M}, \ P_i(X) = C, \ i = 0, \ldots, n-1, \tag{2.284}$$

$$\{x_i\}_{i=0}^{n} \subset X, \ \|x_0\| \le M, \tag{2.285}$$

$$v_i \in \partial_\delta f(x_i) \setminus \{0\}, \ i = 0, 1, \ldots, n-1 \tag{2.286}$$

$$\{\alpha_i\}_{i=0}^{n-1} \subset [\beta_1, \beta_0], \ \{\eta_i\}_{i=0}^{n-1}, \ \{\xi_i\}_{i=0}^{n-1} \subset B_X(0, \delta) \tag{2.287}$$

and that for all integers $i = 0, \ldots, n-1$,

$$x_{i+1} = P_i(x_i - \alpha_i \|v_i\|^{-1} v_i - \alpha_i \xi_i) - \eta_i. \tag{2.288}$$

We claim that $d(x_k, C_{min}) \le \epsilon$ for all integers k satisfying $n_0 \le k \le n$.
Assume that an integer

$$k \in [0, n-1],$$

$$\|x_k\| \le K_* \quad f(x_k) > \inf(f, C) + \bar{\epsilon}/4. \tag{2.289}$$

It follows from (2.282), (2.283), and (2.286)–(2.289) that Lemma 2.24 holds with $\delta_f = \delta, \delta_C = \delta, \epsilon = \bar{\epsilon}/4, K_0 = K_*, \alpha = \alpha_k, x = x_k, v = v_k, \xi = \xi_k + \|v_k\|^1 v_k$, and $y = x_{k+1}$ and this implies that

$$\begin{aligned}
\|x_{k+1} - \bar{x}\|^2 &\le \|x_k - \bar{x}\|^2 - \alpha_k (16\bar{L})^{-1}\bar{\epsilon} \\
&\quad + 2\alpha_k^2 + \delta^2 + 2\delta(K_* + \bar{K} + 2) \\
&\le \|x_k - \bar{x}\|^2 - \alpha_k (16\bar{L})^{-1}\bar{\epsilon} + 2\alpha_k^2 + 2\delta(K_* + \bar{K} + 3).
\end{aligned}$$

Together with (2.278), (2.282), and (2.287), this implies that

$$\begin{aligned}
\|x_{k+1} - \bar{x}\|^2 &\le \|x_k - \bar{x}\|^2 - \alpha_k (32\bar{L})^{-1}\bar{\epsilon} + 2\delta(\bar{K} + 3 + K_*) \\
&\le \|x_k - \bar{x}\|^2 - (32\bar{L})^{-1}\bar{\epsilon}\beta_1 + 2\delta(\bar{K} + 3 + K_*) \\
&\le \|x_k - \bar{x}\|^2 - \beta_1 (64\bar{L})^{-1}\bar{\epsilon}.
\end{aligned}$$

Thus we have shown that the following property holds:

(P2)   if an integer $k \in [0, n-1]$ and (2.289) is valid, then we have

$$\|x_{k+1} - \bar{x}\|^2 \le \|x_k - \bar{x}\|^2 - (64\bar{L})^{-1}\beta_1\bar{\epsilon}.$$

We claim that there exists an integer $j \in \{1, \ldots, n_0\}$ for which

$$f(x_j) \le \inf(f, C) + \bar{\epsilon}/4.$$

Assume the contrary. Then we have

$$f(x_j) > \inf(f, C) + \bar{\epsilon}/4, \quad j = 1, \ldots, n_0. \tag{2.290}$$

It follows from (2.10), (2.283), (2.287), and (2.288) that for all integers $i = 0, \ldots, n-1$, we have

$$\begin{aligned}
\|x_{i+1} - \bar{x}\| &\le 1 + \|x_i - \alpha_i\|v_i\|^{-1}v_i - \alpha_i\xi_i - \bar{x}\| \\
&\le \|x_i - \bar{x}\| + 3.
\end{aligned} \tag{2.291}$$

By (2.281), (2.283), (2.285), and (2.291) for $i = 0, \ldots, n_0$,

$$\|x_i - \bar{x}\| \le \|x_0 - \bar{x}\| + 3i, \tag{2.292}$$

$$\|x_i\| \le 2\|\bar{x}\| + M + 3n_0 < K_*. \tag{2.293}$$

Let

$$k \in \{1, \ldots, n_0 - 1\}. \tag{2.294}$$

It follows from (2.290), (2.293), (2.294), and property (P2) that

$$\|x_{k+1} - \bar{x}\|^2 \le \|x_k - \bar{x}\|^2 - (64\bar{L})^{-1}\beta_1\bar{\epsilon}. \tag{2.295}$$

Relations (2.295), (2.275), (2.283), (2.285), and (2.292) imply that

$$(M + \|\bar{x}\| + 3)^2 \ge \|x_1 - \bar{x}\|^2 - \|x_{n_0} - \bar{x}\|^2$$
$$= \sum_{i=1}^{n_0-1} [\|x_i - \bar{x}\|^2 - \|x_{i+1} - \bar{x}\|^2]$$
$$\ge (n_0 - 1)(64\bar{L})^{-1}\bar{\epsilon}\beta_1 \ge \beta_1 n_0 (128\bar{L})^{-1}\bar{\epsilon},$$
$$(n_0/2)(64\bar{L})^{-1}\bar{\epsilon}\beta_1 \le (2M + 3)^2.$$

This contradicts (2.280). The contradiction we have reached proves that there exists an integer

$$j \in \{1, \ldots, n_0\} \tag{2.296}$$

for which

$$f(x_j) \le \inf(f, C) + \bar{\epsilon}/4. \tag{2.297}$$

By (2.284), (2.287), and (2.288), we have

$$d(x_j, C) \le \delta. \tag{2.298}$$

Relations (2.277), (2.282), (2.297) and (2.298) imply that

$$d(x_j, C_{min}) \le \epsilon. \tag{2.299}$$

We claim that for all integers i satisfying $j \le i \le n$, we have

$$d(x_i, C_{min}) \le \epsilon.$$

Assume the contrary. Then there exists an integer

$$k \in [j, n] \tag{2.300}$$

for which

$$d(x_k, C_{min}) > \epsilon. \tag{2.301}$$

By (2.296) and (2.299)–(2.301),

$$k > j. \tag{2.302}$$

We may assume without loss of generality that

$$d(x_i, C_{min}) \le \epsilon \text{ for all integers } i \text{ satisfying } j \le i < k. \tag{2.303}$$

Then

$$d(x_{k-1}, C_{min}) \le \epsilon. \tag{2.304}$$

There are two cases:

$$f(x_{k-1}) \le \inf(f, C) + \bar{\epsilon}/4; \tag{2.305}$$

$$f(x_{k-1}) > \inf(f, C) + \bar{\epsilon}/4. \tag{2.306}$$

Assume that (2.305) is valid. In view of (2.7), (2.8), (2.275), (2.276), and (2.305),

$$x_{k-1} \in X_0 \subset B_X(0, \bar{K}). \tag{2.307}$$

By (2.284), (2.287), and (2.288), there exists a point $z \in C$ such that

$$\|x_{k-1} - z\| \le \delta. \tag{2.308}$$

It follows from (2.10), (2.287), (2.288), and (2.308) that

$$\begin{aligned}
&\|x_k - z\| \\
&\le \delta + \|z - P_{k-1}(x_{k-1} - \alpha_{k-1}\|v_{k-1}\|^{-1}v_{k-1} - \alpha_{k-1}\xi_{k-1})\| \\
&\le \delta + \|z - x_{k-1}\| + \alpha_{k-1} + \delta < 3\delta + \alpha_{k-1}.
\end{aligned} \tag{2.309}$$

Relations (2.277), (2.282), (2.305), and (2.308) imply that

$$d(x_{k-1}, C_{min}) \le \epsilon/4. \tag{2.310}$$

By (2.276), (2.278), (2.282), (2.287), (2.308), and (2.309),

$$\begin{aligned}
\|x_k - x_{k-1}\| &\le \|x_k - z\| + \|z - x_{k+1}\| \\
&\le 4\delta + \alpha_{k-1} < \bar{\epsilon} < \epsilon/8.
\end{aligned} \tag{2.311}$$

In view of (2.310) and (2.311),

$$d(x_k, C_{min}) \leq \epsilon.$$

This inequality contradicts (2.301). The contradiction we have reached proves (2.306).
 In view of (2.7), (2.8), and (2.304),

$$\|x_{k-1}\| \leq \bar{K} + 1. \tag{2.312}$$

It follows from (2.286)–(2.288), (2.306), and (2.312) that Lemma 2.26 holds with

$$P_C = P_i, \ K_0 = \bar{K} + 1, \ x = x_{k-1}, \ y = x_k, \ v = v_{k-1}, \ \xi = \xi_{k-1} + \|v_{k-1}\|^{-1} v,$$

$$\alpha = \alpha_{k-1}, \ \epsilon = 4^{-1}\bar{\epsilon}, \ \delta_f = \delta, \ \delta_C = \delta$$

, and combining with (2.278), (2.282), (2.287), and (2.304), this implies that

$$
\begin{aligned}
d(x_k, &C_{min})^2 \\
&\leq d(x_{k-1}, C_{min})^2 - \alpha_{k-1}(16\bar{L})^{-1}\bar{\epsilon} + 2\alpha_{k-1}^2 + \delta^2 + 2\delta(2\bar{K} + 4) \\
&\leq d(x_{k-1}, C_{min})^2 - (32\bar{L})^{-1}\alpha_{k-1}\bar{\epsilon} + 2\delta(2\bar{K} + 5) \\
&\leq d(x_{k-1}, C_{min})^2 - (32\bar{L})^{-1}\beta_1\bar{\epsilon} + 2\delta(2\bar{K} + 5) \\
&\leq d(x_{k-1}, C_{min})^2 \leq \epsilon^2.
\end{aligned}
$$

This contradicts (2.301). The contradiction we have reached proves that

$$d(x_i, C_{min}) \leq \epsilon$$

for all integers i satisfying $j \leq i \leq n$. In view of inequality $n_0 \geq j$, Theorem 2.12 is proved.

2.15 Proof of Theorem 2.13

We may assume without loss of generality that

$$M > \bar{K} + 4, \ \epsilon < 1. \tag{2.313}$$

. There exists $L_0 > \bar{L}$ such that

$$|f(z_1) - f(z_2)| \leq L_0 \|z_1 - z_2\| \text{ for all } z_1, z_2 \in B_X(0, M + 4). \tag{2.314}$$

In view of Proposition 2.28, there exist a number

$$\bar{\epsilon} \in (0, \epsilon/8) \tag{2.315}$$

such that

$$\text{if } x \in X, \ d(x, C) \le 2\bar{\epsilon}, \ \text{and } f(x) \le \inf(f, C) + 2\bar{\epsilon},$$

$$\text{then } d(x, C_{min}) \le \epsilon. \tag{2.316}$$

Fix

$$\bar{x} \in C_{min} \tag{2.317}$$

and

$$\epsilon_0 \in (0, 4^{-1}\bar{\epsilon}). \tag{2.318}$$

Since $\lim_{i \to \infty} \alpha_i = 0$ (see (2.83)), there is an integer $p_0 > 0$ such that

$$M + 4 < p_0 \tag{2.319}$$

and that for all integers $p \ge p_0 - 1$, we have

$$\alpha_p < (20(\bar{L} + L_0 + 2))^{-2} \epsilon_0. \tag{2.320}$$

Since $\sum_{i=0}^{\infty} \alpha_i = \infty$ (see (2.83)), there exist a natural number

$$n_0 > p_0 + 4 \tag{2.321}$$

such that

$$\sum_{i=p_0}^{n_0-1} \alpha_i > (4p_0 + M + \|\bar{x}\|)^2 \epsilon_0^{-1} 16\bar{L}. \tag{2.322}$$

Fix

$$K_* > \bar{K} + 4 + M + 4n_0 + 4\|\bar{x}\| + 5L_0 + 5\bar{L} \tag{2.323}$$

and a positive number δ such that

$$6\delta(K_* + 1) < 16^{-1}\epsilon_0. \tag{2.324}$$

Assume that an integer $n \ge n_0$ and that

$$\{P_i\}_{i=0}^{n-1} \subset \mathcal{M}, \ \{x_i\}_{i=0}^n \subset X,$$

$$P_i(X) = C, \ i = 0, \dots, n-1, \tag{2.325}$$

$$\|x_0\| \le M, \tag{2.326}$$

$\{\xi_i\}_{i=0}^{n-1} \subset X$ and that for all $i = 0, \dots, n-1$,

$$B_X(\xi_i, \delta) \cap \partial_\delta f(x_i) \ne \emptyset, \tag{2.327}$$

$$\|x_{i+1} - P_i(x_i - \alpha_i \xi_i)\| \le \alpha_i \delta. \tag{2.328}$$

In order to prove the theorem, it is sufficient to show that

$$d(x_k, C_{min}) \le \epsilon \text{ for all integers } k \text{ satisfying } n_0 \le k \le n.$$

By (2.83), (2.324)–(2.326), and (2.328),

$$\|x_i\| \le M + 1 \text{ for all integers } i = 0, \dots, n. \tag{2.329}$$

Assume that an integer

$$k \in [p_0, n-1], \tag{2.330}$$

$$f(x_k) > \inf(f, C) + \epsilon_0. \tag{2.331}$$

In view of (2.314), (2.317), (2.323), (2.324), (2.326)–(2.328), and (2.331), the conditions of Lemma 2.21 hold with $K_0 = M, \delta_f = \delta, \delta_C = \alpha_k \delta, \epsilon = \epsilon_0, \alpha = \alpha_k,$ $x = x_k, \xi = \xi_k,$ and $y = x_{k+1}$, and combined with (2.320), (2.323), (2.324), and (2.330), this Lemma implies that

$$\begin{aligned}
\|x_{k+1} - \bar{x}\|^2 &\le \|x_k - \bar{x}\|^2 - 2^{-1}\alpha_k \epsilon_0 \\
&\quad + \alpha_k^2 \delta^2 + 2\delta\alpha_k(M + \bar{K} + 5L_0 + 5\bar{L}) + 25\alpha_k^2(L_0 + \bar{L})^2 \\
&\le \|x_k - \bar{x}\|^2 - 4^{-1}\alpha_k \epsilon_0 \\
&\quad + 2\delta\alpha_k(M + \bar{K} + 5L_0 + 5\bar{L}) + 25\alpha_k^2(L_0 + \bar{L})^2 \\
&\le \|x_k - \bar{x}\|^2 - 8^{-1}\alpha_k \epsilon_0 + 2\delta\alpha_k(M + \bar{K} + 5L_0 + 5\bar{L}) \\
&\le \|x_k - \bar{x}\|^2 - 16^{-1}\alpha_k \epsilon_0.
\end{aligned}$$

Thus we have shown that the following property holds:

(P3)   If an integer $k \in [p_0, n-1]$ and (2.331) is valid, then we have

$$\|x_{k+1} - \bar{x}\|^2 \leq \|x_k - \bar{x}\|^2 - 16^{-1}\alpha_k\epsilon_0.$$

We claim that there exists an integer $j \in \{p_0, \ldots, n_0\}$ such that

$$f(x_j) \leq \inf(f, C) + \epsilon_0.$$

Assume the contrary. Then

$$f(x_i) > \inf(f, C) + \epsilon_0, \quad i = p_0, \ldots, n_0. \tag{2.332}$$

It follows from (2.332) and property (P3) that

$$\|x_{i+1} - \bar{x}\|^2 \leq \|x_i - \bar{x}\|^2 - 16^{-1}\alpha_i\epsilon_0. \tag{2.333}$$

Relations (2.83), (2.313), (2.317), (2.325), (2.328), and (2.333) imply that

$$(2M + 1)^2 \geq \|x_{p_0} - \bar{x}\|^2 - \|x_{n_0} - \bar{x}\|^2$$

$$= \sum_{i=p_0}^{n_0-1} [\|x_i - \bar{x}\|^2 - \|x_{i+1} - \bar{x}\|^2] \geq 16^{-1}\epsilon_0 \sum_{i=p_0}^{n_0-1} \alpha_i$$

and

$$\sum_{i=p_0}^{n_0-1} \alpha_i \leq 16\epsilon_0^{-1}(2M + 1)^2.$$

This contradicts (2.322). The contradiction we have reached proves that there exists an integer

$$j \in \{p_0, \ldots, n_0\} \tag{2.334}$$

such that

$$f(x_j) \leq \inf(f, C) + \epsilon_0. \tag{2.335}$$

By (2.318), (2.324), and (2.328), we have

$$d(x_j, C) \leq \alpha_{j-1}\delta < \bar{\epsilon}. \tag{2.336}$$

In view of (2.316), (2.335), and (2.336),

$$d(x_j, C_{min}) \leq \epsilon. \tag{2.337}$$

We claim that for all integers i satisfying $j \leq i \leq n$,

$$d(x_i, C_{min}) \le \epsilon.$$

Assume the contrary. Then there exists an integer

$$k \in [j, n] \tag{2.338}$$

for which

$$d(x_k, C_{min}) > \epsilon. \tag{2.339}$$

By (2.337)–(2.339), we have

$$k > j \ge p_0. \tag{2.340}$$

By (2.337)–(2.340), we may assume without loss of generality that

$$d(x_i, C_{min}) \le \epsilon \text{ for all integers } i \text{ satisfying } j \le i < k. \tag{2.341}$$

Thus in view of (2.341),

$$d(x_{k-1}, C_{min}) \le \epsilon. \tag{2.342}$$

There are two cases:

$$f(x_{k-1}) \le \inf(f, C) + \epsilon_0; \tag{2.343}$$

$$f(x_{k-1}) > \inf(f, C) + \epsilon_0. \tag{2.344}$$

Assume that (2.343) is valid. By (2.325) and (2.328), there exists a point

$$z \in C \tag{2.345}$$

such that

$$\|x_{k-1} - z\| \le \delta \alpha_{k-2}. \tag{2.346}$$

By (2.10), (2.328), (2.345), and (2.346),

$$\begin{aligned}
\|x_k - z\| &\le \alpha_{k-1}\delta + \|z - P_{k-1}(x_{k-1} - \alpha_{k-1}\xi_{k-1})\| \\
&\le \delta\alpha_{k-1} + \|z - x_{k-1}\| + \alpha_{k-1}\|\xi_{k-1}\| \\
&\le \alpha_{k-1}\|\xi_{k-1}\| + \delta(\alpha_{k-1} + \alpha_{k-2}).
\end{aligned} \tag{2.347}$$

Lemma 2.19, (2.83), (2.313), (2.314), (2.345), and (2.346) imply that

$$\partial_\delta f(x_{k-1}) \subset B_X(0, L_0 + \delta) \subset B_X(0, L_0 + 1). \tag{2.348}$$

In view of (2.327) and (2.348),

$$\|\xi_{k-1}\| \leq L_0 + 2. \tag{2.349}$$

Equation (2.349) implies that

$$\alpha_{k-1}\|\xi_{k-1}\| \leq \alpha_{k-1}(L_0 + 2). \tag{2.350}$$

By (2.347) and (2.350),

$$\|x_k - z\| \leq \delta(\alpha_{k-1} + \alpha_{k-2}) + \alpha_{k-1}(L_0 + 2). \tag{2.351}$$

It follows from (2.320), (2.324), (2.334), (2.338), (2.345), and (2.351) that

$$d(x_k, C) \leq \delta(\alpha_{k-1} + \alpha_{k-2}) + \alpha_{k-1}(L_0 + 2) \leq \epsilon_0. \tag{2.352}$$

In view of (2.346) and (2.351),

$$\begin{aligned}
\|x_k - x_{k-1}\| &\leq \|x_k - z\| + \|z - x_{k-1}\| \\
&\leq \delta\alpha_{k-2} + \delta(\alpha_{k-1} + \alpha_{k-2}) + \alpha_{k-1}(L_0 + 2).
\end{aligned} \tag{2.353}$$

It follows from (2.83), (2.314), (2.346), (2.352), and (2.353) that

$$|f(x_{k-1}) - f(x_k)| \leq L_0\|x_{k-1} - x_k\| \leq 2L_0\delta(\alpha_{k-1} + \alpha_{k-2}) + \alpha_{k-1}L_0(L_0 + 2). \tag{2.354}$$

In view of (2.320), (2.324), (2.334), (2.340), (2.343), and (2.354), we have

$$\begin{aligned}
f(x_k) &\leq f(x_{k-1}) + 2L_0\delta(\alpha_{k-1} + \alpha_{k-2}) + \alpha_{k-1}L_0(L_0 + 2) \\
&\leq \inf(f, C) + \epsilon_0 + 8^{-1}\epsilon_0 + 20^{-1}\epsilon_0 \leq \inf(f, C) + 2\epsilon_0. \quad (2.355)
\end{aligned}$$

It follows from (2.316), (2.318), (2.352), and (2.355) that

$$d(x_k, C_{min}) \leq \epsilon.$$

This inequality contradicts (2.339). The contradiction we have reached proves (2.344).

It follows from (2.83), (2.314), (2.318), (2.323)–(2.325), (2.328), and (2.344) that Lemma 2.22 holds with

$$x = x_{k-1}, \quad y = x_k, \quad \xi = \xi_{k-1}, \quad \alpha = \alpha_{k-1}, \quad K_0 = M + 1,$$

$$\epsilon = \epsilon_0, \quad \delta_f = \delta, \quad \delta_C = \alpha_{k-1}\delta$$

, and combined with (2.320), (2.323), (2.324), (2.334), (2.340), and (2.342), this implies that

$$
\begin{aligned}
d(x_k, C_{min})^2 &\leq d(x_{k-1}, C_{min})^2 - 2^{-1}\alpha_{k-1}\epsilon_0 + 25\alpha_{k-1}^2(L_0 + \bar{L})^2 \\
&\quad + \alpha_{k-1}^2\delta^2 + 2\alpha_{k-1}\delta(M + 1 + \bar{K} + 5L_0 + 5\bar{L}) \\
&\leq d(x_{k-1}, C_{min})^2 - 4^{-1}\alpha_{k-1}\epsilon_0 + 2\alpha_{k-1}\delta(M + 2 + 2\bar{K} + 5L_0 + 5\bar{L}) \\
&\leq d(x_{k-1}, C_{min})^2 - 8^{-1}\alpha_{k-1}\epsilon_0 < d(x_{k-1}, C_{min})^2 \leq \epsilon^2
\end{aligned}
$$

and

$$
d(x_k, C_{min}) \leq \epsilon.
$$

This contradicts (2.339). The contradiction we have reached proves that $d(x_i, C_{min}) \leq \epsilon$ for all integers i satisfying $j \leq i \leq n$. This completes the proof of Theorem 2.13.

2.16   Proof of Theorem 2.14

Fix

$$
\bar{x} \in C_{min} \tag{2.356}
$$

In view of Proposition 2.28, there exist a number

$$
\bar{\epsilon} \in (0, \epsilon/8) \tag{2.357}
$$

such that

$$
\text{if } x \in X, \ d(x, C) \leq 2\bar{\epsilon} \text{ and } f(x) \leq \inf(f, C) + 2\bar{\epsilon},
$$

$$
\text{then } d(x, C_{min}) \leq \epsilon. \tag{2.358}
$$

Fix

$$
\epsilon_0 \in (0, 4^{-1}\bar{\epsilon}). \tag{2.359}
$$

Since $\lim_{i \to \infty} \alpha_i = 0$ (see (2.88)), there is an integer $p_0 > 0$ such that

$$
M + 4 < p_0 \tag{2.360}
$$

and that for all integers $p \geq p_0 - 1$, we have

$$\alpha_p < (20(\bar{L} + L_0 + 2))^{-2}\epsilon_0. \tag{2.361}$$

Since $\sum_{i=0}^{\infty} \alpha_i = \infty$ (see (2.88)), there exist a natural number

$$n_0 > p_0 + 4 \tag{2.362}$$

such that

$$\sum_{i=p_0}^{n_0-1} \alpha_i > 16(4p_0 + 2M + \|\bar{x}\| + 2)^2 \epsilon_0^{-1}. \tag{2.363}$$

Fix

$$K_* > \bar{K} + 4 + 3M + 4n_0 + 4\|\bar{x}\| + 5L_0 + 5\bar{L} \tag{2.364}$$

and a positive number δ such that

$$6\delta(K_* + 1) < 16^{-1}\epsilon_0. \tag{2.365}$$

Assume that an integer $n \geq n_0$ and that

$$\{P_i\}_{i=0}^{n-1} \subset \mathcal{M}, \ \{x_i\}_{i=0}^{n} \subset X,$$

$$P_i(X) = C, \ i = 0, \dots, n - 1, \tag{2.366}$$

$$\|x_0\| \leq M, \tag{2.367}$$

$\{\xi_i\}_{i=0}^{n-1} \subset X$ and that for all $i = 0, \dots, n - 1$,

$$B_X(\xi_i, \delta) \cap \partial_\delta f(x_i) \neq \emptyset, \tag{2.368}$$

$$\|x_{i+1} - P_i(x_i - \alpha_i \xi_i)\| \leq \alpha_i \delta \tag{2.369}$$

and (2.88) is true. In order to prove the theorem, it is sufficient to show that

$$d(x_k, C_{min}) \leq \epsilon \text{ for all integers } k \text{ satisfying } n_0 \leq k \leq n.$$

We show that for all $t = 0, , \dots, n$,

$$\|x_t - \bar{x}\| \leq 2 + M + \bar{K}. \tag{2.370}$$

(In view of (2.356) and (2.367), the inequality above holds with $t = 0$.) Assume that an integer t satisfies $0 \leq t < n$ and that (2.370) is true. We show that our assumption holds for $t + 1$ too. There are two cases:

$$f(x_t) \leq \inf(f, C) + 4; \tag{2.371}$$

$$f(x_t) > \inf(f, C) + 4. \tag{2.372}$$

Assume that (2.371) holds. In view of (2.7), (2.8), and (2.371),

$$\|x_t\| \leq \bar{K}. \tag{2.373}$$

Lemma 2.19, (2.9), and (2.373) imply that

$$\partial_\delta f(x_t) \subset B_X(0, \bar{L} + 1). \tag{2.374}$$

By (2.365), (2.368), and (2.374),

$$\|\xi_t\| \leq \bar{L} + 2. \tag{2.375}$$

It follows from (2.7), (2.8), (2.10), (2.88), (2.356), (2.365), (2.366), (2.369), (2.373), and (2.375) that

$$\|x_{t+1} - \bar{x}\| \leq \|x_{t+1} - P_t(x_t - \alpha_t \xi_t)\| + \|P_t(x_t - \alpha_t \xi_t) - \bar{x}\|$$

$$\leq \alpha_t \delta + \|x_t - \alpha_t \xi_t - \bar{x}\| \leq \alpha_t \delta + \|x_t - \bar{x}\| + \alpha_t \|\xi_t\|$$

$$\leq 1 + 2\bar{K} + \alpha \|\xi_t\| \leq 3 + 2\bar{K} \leq M + 2 + \bar{K}.$$

Thus

$$\|x_{t+1} - \bar{x}\| \leq M + 2 + \bar{K}. \tag{2.376}$$

Assume that (2.372) holds. In view of (2.356), (2.364), (2.365), (2.368), (2.369), and (2.372), we apply Lemma 2.21 with

$$P_C = P_t, \ \delta_f = \delta, \ \delta_C = \alpha_t \delta,$$

$$K_0 = M + 4 + \bar{K}, \ \alpha = \alpha_t, \ \xi = \xi_t, \ x = x_t, \ y = x_{t+1}, \epsilon = 4$$

, and this Lemma together with (2.88), (2.364), (2.365), and (2.370) implies that

$$\|x_{t+1} - \bar{x}\|^2 \leq \|x_t - \bar{x}\|^2 - 2\alpha_t + \alpha_t^2 \delta^2$$

$$+ 2\alpha_t \delta(2\bar{K} + M + 2 + 5L_0 + 5\bar{L}) + 25\alpha_t^2(L_0 + \bar{L})^2$$

$$\leq \|x_t - \bar{x}\|^2 - \alpha_t + 2\alpha_t \delta(2\bar{K} + M + 3 + 5L_0 + 5\bar{L})$$

$$\leq \|x_t - \bar{x}\|^2$$

and

$$\|x_{t+1} - \bar{x}\| \le \|x_t - \bar{x}\| \le \bar{K} + M + 2.$$

Thus (2.376) holds in the both cases. Therefore by induction we showed that (2.370) holds for all $t = 0, \dots, n$.

Assume that an integer

$$k \in [p_0, n - 1], \tag{2.377}$$

$$f(x_k) > \inf(f, C) + \epsilon_0. \tag{2.378}$$

In view of (2.356), (2.359), (2.364), (2.365), (2.368)–(2.370), (2.377), and (2.378), the conditions of Lemma 2.21 hold with $K_0 = 2M + 4$, $\delta_f = \delta$, $\delta_C = \alpha_k \delta$, $\epsilon = \epsilon_0$, $\alpha = \alpha_k$, $x = x_k$, $\xi = \xi_k$, and $y = x_{k+1}$, and combined with (2.361), (2.365), and (2.377), this Lemma implies that

$$
\begin{aligned}
\|x_{k+1} - \bar{x}\|^2 &\le \|x_k - \bar{x}\|^2 - 2^{-1}\alpha_k \epsilon_0 \\
&\quad + \alpha_k^2 \delta^2 + 2\delta\alpha_k(2M + 2\bar{K} + 5L_0 + 5\bar{L} + 4) + 25\alpha_k^2(L_0 + \bar{L})^2 \\
&\le \|x_k - \bar{x}\|^2 - 4^{-1}\alpha_k \epsilon_0 \\
&\quad + 2\delta\alpha_k(2M + 2\bar{K} + 5L_0 + 5\bar{L} + 4) \\
&\le \|x_k - \bar{x}\|^2 - 8^{-1}\alpha_k \epsilon_0.
\end{aligned}
$$

Thus we have shown that the following property holds:

(P4)   If an integer $k \in [p_0, n - 1]$ and (2.378) is valid, then we have

$$\|x_{k+1} - \bar{x}\|^2 \le \|x_k - \bar{x}\|^2 - 8^{-1}\alpha_k \epsilon_0.$$

We claim that there exists an integer $j \in \{p_0, \dots, n_0\}$ such that

$$f(x_j) \le \inf(f, C) + \epsilon_0.$$

Assume the contrary. Then

$$f(x_i) > \inf(f, C) + \epsilon_0, \quad i = p_0, \dots, n_0. \tag{2.379}$$

It follows from (2.379) and property (P4) that for all $i = p_0, \dots, n_0 - 1$,

$$\|x_{i+1} - \bar{x}\|^2 \le \|x_i - \bar{x}\|^2 - 8^{-1}\alpha_i \epsilon_0. \tag{2.380}$$

Relations (2.370) and (2.380) imply that

$$(M + \bar{K} + 2)^2 \geq \|x_{p_0} - \bar{x}\|^2 - \|x_{n_0} - \bar{x}\|^2$$

$$= \sum_{i=p_0}^{n_0-1} [\|x_i - \bar{x}\|^2 - \|x_{i+1} - \bar{x}\|^2] \geq 8^{-1}\epsilon_0 \sum_{i=p_0}^{n_0-1} \alpha_i$$

and

$$\sum_{i=p_0}^{n_0-1} \alpha_i \leq 8\epsilon_0^{-1}(M + \bar{K} + 2)^2.$$

This contradicts (2.363). The contradiction we have reached proves that there exists an integer

$$j \in \{p_0, \ldots, n_0\} \tag{2.381}$$

such that

$$f(x_j) \leq \inf(f, C) + \epsilon_0. \tag{2.382}$$

By (2.365), (2.366), and (2.369), we have

$$d(x_j, C) \leq \alpha_{j-1}\delta < \bar{\epsilon}. \tag{2.383}$$

In view of (2.358), (2.359), (2.382), and (2.383),

$$d(x_j, C_{min}) \leq \epsilon. \tag{2.384}$$

We claim that for all integers i satisfying $j \leq i \leq n$,

$$d(x_i, C_{min}) \leq \epsilon.$$

Assume the contrary. Then there exists an integer

$$k \in [j, n] \tag{2.385}$$

for which

$$d(x_k, C_{min}) > \epsilon. \tag{2.386}$$

By (2.384)–(2.386), we have

$$k > j \geq p_0. \tag{2.387}$$

We may assume without loss of generality that

$$d(x_i, C_{min}) \leq \epsilon \text{ for all integers } i \text{ satisfying } j \leq i < k. \tag{2.388}$$

Thus in view of (2.388),

$$d(x_{k-1}, C_{min}) \leq \epsilon. \tag{2.389}$$

There are two cases:

$$f(x_{k-1}) \leq \inf(f, C) + \epsilon_0; \tag{2.390}$$

$$f(x_{k-1}) > \inf(f, C) + \epsilon_0. \tag{2.391}$$

Assume that (2.390) is valid. By (2.366) and (2.369), there exists a point $z \in C$ such that

$$\|x_{k-1} - z\| \leq \delta\alpha_{k-2}. \tag{2.392}$$

By (2.10), (2.366), (2.369), and (2.392),

$$\begin{aligned}
\|x_k - z\| &\leq \alpha_{k-1}\delta + \|z - P_{k-1}(x_{k-1} - \alpha_{k-1}\xi_{k-1})\| \\
&\leq \delta\alpha_{k-1} + \|z - x_{k-1}\| + \alpha_{k-1}\|\xi_{k-1}\| \\
&\leq \alpha_{k-1}\|\xi_{k-1}\| + \delta(\alpha_{k-1} + \alpha_{k-2}).
\end{aligned} \tag{2.393}$$

Lemma 2.19, (2.87), and (2.370) imply that

$$\partial_\delta f(x_{k-1}) \subset B_X(0, L_0 + \delta) \subset B_X(0, L_0 + 1). \tag{2.394}$$

In view of (2.368) and (2.394),

$$\|\xi_{k-1}\| \leq L_0 + 2.$$

By the relation above,

$$\alpha_{k-1}\|\xi_{k-1}\| \leq \alpha_{k-1}(L_0 + 2).$$

Together with (2.393), this implies that

$$\|x_k - z\| \leq \delta(\alpha_{k-1} + \alpha_{k-2}) + \alpha_{k-1}(L_0 + 2). \tag{2.395}$$

It follows from the inclusion $z \in C$, (2.361), (2.365), and (2.387) that

$$d(x_k, C) \leq \delta(\alpha_{k-1} + \alpha_{k-2}) + \alpha_{k-1}(L_0 + 2) \leq \epsilon_0. \tag{2.396}$$

In view of (2.392) and (2.395),

$$\|x_k - x_{k-1}\| \le \|x_k - z\| + \|z - x_{k-1}\|$$
$$\le \delta\alpha_{k-2} + \delta(\alpha_{k-1} + \alpha_{k-2}) + \alpha_{k-1}(L_0 + 2). \tag{2.397}$$

It follows from (2.87), (2.370), and (2.397) that

$$|f(x_{k-1}) - f(x_k)| \le L_0 \|x_{k-1} - x_k\|$$
$$\le 2L_0\delta(\alpha_{k-1} + \alpha_{k-2}) + \alpha_{k-1}L_0(L_0 + 2). \tag{2.398}$$

In view of (2.361), (2.364), (2.365), (2.387), (2.390), and (2.398), we have

$$f(x_k) \le f(x_{k-1}) + 2L_0\delta(\alpha_{k-1} + \alpha_{k-2}) + \alpha_{k-1}L_0(L_0 + 2)$$
$$\le \inf(f, C) + \epsilon_0 + 8^{-1}\epsilon_0 + 20^{-1}\epsilon_0 \le \inf(f, C) + 2\epsilon_0. \tag{2.399}$$

It follows from (2.358), (2.359), (2.396), and (2.399) that

$$d(x_k, C_{min}) \le \epsilon.$$

This inequality contradicts (2.386). The contradiction we have reached proves (2.391).

It follows from (2.87), (2.356), (2.365), (2.368)–(2.370), and (2.391) that Lemma 2.22 holds with

$$x = x_{k-1}, \ y = x_k, \ \xi = \xi_{k-1}, \ \alpha = \alpha_{k-1}, \ K_0 = M + 2\bar{K} + 2,$$

$$\epsilon = \epsilon_0, \ \delta_f = \delta, \ \delta_C = \alpha_{k-1}\delta$$

, and combined with (2.361), (2.364), (2.365), (2.387), and (2.389) this implies that

$$d(x_k, C_{min})^2 \le d(x_{k-1}, C_{min})^2 - 2^{-1}\alpha_{k-1}\epsilon_0 + 25\alpha_{k-1}^2(L_0 + \bar{L})^2$$
$$+ \alpha_{k-1}^2\delta^2 + 2\alpha_{k-1}\delta(M + 2 + 2\bar{K} + 5L_0 + 5\bar{L})$$
$$\le d(x_{k-1}, C_{min})^2 - 4^{-1}\alpha_{k-1}\epsilon_0 + 2\alpha_{k-1}\delta(M + 3 + 2\bar{K} + 5L_0 + 5\bar{L})$$
$$\le d(x_{k-1}, C_{min})^2 - 8^{-1}\alpha_{k-1}\epsilon_0$$

and

$$d(x_k, C_{min}) \le d(x_{k-1}, C_{min}) \le \epsilon.$$

This contradicts (2.386). The contradiction we have reached proves that $d(x_i, C_{min}) \leq \epsilon$ for all integers i satisfying $j \leq i \leq n$. This completes the proof of Theorem 2.14.

2.17 Proof of Theorem 2.15

We may assume that without loss of generality

$$\epsilon < 1, \ M > \bar{K} + 4. \tag{2.400}$$

There exists $L_0 > \bar{L}$ such that

$$|f(z_1) - f(x_2)| \leq L_0 \|z_1 - z_2\| \text{ for all } z_1, z_2 \in B_X(0, 3M + 4). \tag{2.401}$$

Proposition 2.28 implies that there exists

$$\bar{\epsilon} \in (0, \epsilon/8) \tag{2.402}$$

such that

$$\text{if } x \in X, \ d(x, C) \leq 2\bar{\epsilon}, and f(x) \leq \inf(f, C) + 2\bar{\epsilon},$$

$$\text{then } d(x, C_{min}) \leq \epsilon. \tag{2.403}$$

Set

$$\beta_0 = (64(L_0 + \bar{L}))^{-2}\bar{\epsilon}. \tag{2.404}$$

Let

$$\beta_1 \in (0, \beta_0). \tag{2.405}$$

There exists an integer $n_0 \geq 4$ such that

$$\beta_1 n_0 > 16^2(3 + 2M)^2 \bar{\epsilon}^{-1} L_0. \tag{2.406}$$

Fix

$$K_* > 6M + 4 + 4n_0 + 5L_0 + 5\bar{L} \tag{2.407}$$

and a positive number δ such that

$$6\delta K_* < (64L_0)^{-1}\bar{\epsilon}\beta_1. \tag{2.408}$$

Assume that an integer $n \geq n_0$,

$$\{P_i\}_{i=0}^{n-1} \subset \mathcal{M}, \ P_i(X) = C, \ i = 0, \ldots, n-1, \tag{2.409}$$

$$\{x_i\}_{i=0}^{n} \subset X, \ \|x_0\| \leq M, \tag{2.410}$$

$\{\xi_i\}_{i=0}^{n-1} \subset X,$

$$\{\alpha_i\}_{i=0}^{n-1} \subset [\beta_1, \beta_0] \tag{2.411}$$

and that for all integers $i = 0, \ldots, n-1$,

$$B_X(\xi_i, \delta) \cap \partial_\delta f(x_i) \neq \emptyset, \tag{2.412}$$

$$\|x_{i+1} - P_i(x_i - \alpha_i \xi_i)\| \leq \delta. \tag{2.413}$$

In order to prove the theorem, it is sufficient to show that

$$d(x_k, C_{min}) \leq \epsilon \text{ for all integers } k \text{ satisfying } n_0 \leq k \leq n.$$

Fix a point

$$\bar{x} \in C_{min}. \tag{2.414}$$

We show that for all $t = 0, , \ldots, n$,

$$\|x_t - \bar{x}\| \leq 2 + M + \bar{K}. \tag{2.415}$$

In view of (2.7), (2.8), (2.410), and (2.414), (2.415) holds with $t = 0$. Assume that an integer t satisfies $0 \leq t < n$ and that (2.415) is true. There are two cases:

$$f(x_t) \leq \inf(f, C) + 4; \tag{2.416}$$

$$f(x_t) > \inf(f, C) + 4. \tag{2.417}$$

Assume that (2.416) holds. In view of (2.7), (2.8), and (2.416),

$$\|x_t\| \leq \bar{K}. \tag{2.418}$$

Lemma 2.19, (2.401) and (2.418) imply that

$$\partial_\delta f(x_t) \subset B_X(0, \bar{L} + 1). \tag{2.419}$$

By (2.412) and (2.419),

$$\|\xi_t\| \le \bar{L} + 2. \tag{2.420}$$

It follows from (2.7), (2.8), (2.10), (2.404), (2.411), (2.413), (2.414), (2.418), and (2.420) that

$$\|x_{t+1} - \bar{x}\| \le \|x_{t+1} - P_t(x_t - \alpha_t \xi_t)\| + \|P_t(x_t - \alpha_t \xi_t) - \bar{x}\|$$
$$\le \delta + \|x_t - \alpha_t \xi_t - \bar{x}\| \le \delta + \|x_t - \bar{x}\| + \alpha_t \|\xi_t\|$$
$$\le 1 + 2\bar{K} + \alpha_t \|\xi_t\| \le 1 + 2\bar{K} + \beta_0 \|\xi_t\| \le 3 + 2\bar{K} \le M + \bar{K} + 2.$$

Thus

$$\|x_{t+1} - \bar{x}\| \le M + 2 + \bar{K}.$$

Assume that (2.417) holds. In view of (2.7), (2.8), (2.407), (2.408), (2.412)–(2.415), and (2.417), Lemma 2.21 holds

$$P_C = P_t, \ \delta_f = \delta, \ \delta_C = \delta,$$

$$K_0 = M + 2\bar{K} + 2, \ \alpha = \alpha_t, \ \xi = \xi_t, \ x = x_t, \ y = x_{t+1}, \epsilon = 4$$

, and this Lemma together with (2.48), (2.404), (2.407), (2.408), and (2.411) implies that

$$\|x_{t+1} - \bar{x}\|^2 \le \|x_t - \bar{x}\|^2 - 2\alpha_t + \delta^2$$
$$+ 2\delta(3\bar{K} + M + 2 + 5L_0 + 5\bar{L}) + 25\alpha_t^2(L_0 + \bar{L})^2$$
$$\le \|x_t - \bar{x}\|^2 - \alpha_t + \delta(3\bar{K} + M + 3 + 5L_0 + 5\bar{L})$$
$$\le \|x_t - \bar{x}\|^2 - \beta_1 + \delta(2\bar{K} + 2M + 3 + 5L_0 + 5\bar{L})$$
$$\le \|x_t - \bar{x}\|^2$$

and

$$\|x_{t+1} - \bar{x}\| \le \bar{K} + M + 2$$

in the both cases. Therefore by induction we showed that (2.415) holds for all $t = 0, \ldots, n$.

Assume that an integer

$$k \in [0, n-1], \tag{2.421}$$

$$f(x_k) > \inf(f, C) + \bar{\epsilon}. \tag{2.422}$$

It follows from (2.401), (2.404), (2.407), (2.408), (2.412)–(2.415), and (2.422) that Lemma 2.21 holds with $\delta_f = \delta, \delta_C = \delta, \epsilon = \bar{\epsilon}, K_0 = \bar{K} + 2M + 2, x = x_k, \xi = \xi_k,$ and $y = x_{k+1}$, and this Lemma together with (2.411) implies that

$$
\begin{aligned}
\|x_{k+1} - \bar{x}\|^2 &\le \|x_k - \bar{x}\|^2 - 2^{-1}\alpha_k\bar{\epsilon} \\
&\quad + \delta^2 + 2\delta(2\bar{K} + 2M + 2 + 5L_0 + \bar{L}) + 25\alpha_k^2(L_0 + \bar{L})^2 \\
&\le \|x_k - \bar{x}\|^2 - 4^{-1}\alpha_k\bar{\epsilon} + 2\delta(2\bar{K} + 2M + 3 + 5L_0 + 5\bar{L}) \\
&\le \|x_k - \bar{x}\|^2 - 4^{-1}\beta_1\bar{\epsilon} + 2\delta(2\bar{K} + 2M + 3 + 5L_0 + 5\bar{L}) \\
&\le \|x_k - \bar{x}\|^2 - 8^{-1}\beta_1\bar{\epsilon}.
\end{aligned}
$$

Thus we have shown that the following property holds:

(P5) if an integer $k \in [0, n - 1]$ and (2.422) is valid, then we have

$$\|x_{k+1} - \bar{x}\|^2 \le \|x_k - \bar{x}\|^2 - 8^{-1}\beta_1\bar{\epsilon}.$$

We claim that there exists an integer $j \in \{1, \ldots, n_0\}$ for which

$$f(x_j) \le \inf(f, C) + \bar{\epsilon}.$$

Assume the contrary. Then we have

$$f(x_j) > \inf(f, C) + \bar{\epsilon}, \quad j = 1, \ldots, n_0. \tag{2.423}$$

It follows from (2.423) and property (P5) that for all $k = 0, \ldots, n_0 - 1$,

$$\|x_{k+1} - \bar{x}\|^2 \le \|x_k - \bar{x}\|^2 - 8^{-1}\beta_1\bar{\epsilon}. \tag{2.424}$$

Relations (2.7), (2.8), (2.410), (2.414), and (2.424) imply that

$$
\begin{aligned}
(M + \bar{K})^2 &\ge \|x_0 - \bar{x}\|^2 - \|x_{n_0} - \bar{x}\|^2 \\
&= \sum_{i=0}^{n_0-1}[\|x_i - \bar{x}\|^2 - \|x_{i+1} - \bar{x}\|^2] \ge 8^{-1}n_0\bar{\epsilon}\beta_1, \\
n_0 &\le 8(M + \bar{K})^2\bar{\epsilon}^{-1}\beta_1^{-1}.
\end{aligned}
$$

This contradicts (2.406). The contradiction we have reached proves that there exists an integer

$$j \in \{p_0, \ldots, n_0\}$$

for which

$$f(x_j) \leq \inf(f, C) + \bar{\epsilon}. \tag{2.425}$$

By (2.408), (2.409), and (2.413), we have

$$d(x_j, C) \leq \delta < \bar{\epsilon}. \tag{2.426}$$

Relations (2.403), (2.425), and (2.426) imply that

$$d(x_j, C_{min}) \leq \epsilon. \tag{2.427}$$

We claim that for all integers i satisfying $j \leq i \leq n$, we have

$$d(x_i, C_{min}) \leq \epsilon.$$

Assume the contrary. Then there exists an integer

$$k \in [j, n] \tag{2.428}$$

for which

$$d(x_k, C_{min}) > \epsilon. \tag{2.429}$$

By (2.426), (2.428), and(2.429),

$$k > j. \tag{2.430}$$

In view of (2.426) and (2.429), we may assume without loss of generality that

$$d(x_i, C_{min}) \leq \epsilon \text{ for all integers } i \text{ satisfying } j \leq i < k. \tag{2.431}$$

In particular

$$d(x_{k-1}, C_{min}) \leq \epsilon. \tag{2.432}$$

There are two cases:

$$f(x_{k-1}) \leq \inf(f, C) + \bar{\epsilon}; \tag{2.433}$$

$$f(x_{k-1}) > \inf(f, C) + \bar{\epsilon}. \tag{2.434}$$

Assume that (2.433) is valid. In view of (2.409) and (2.413), there exists a point

$$z \in C \tag{2.435}$$

such that

$$\|x_{k-1} - z\| \le \delta. \tag{2.436}$$

It follows from (2.10), (2.411), (2.413), (2.435), and (2.436) that

$$
\begin{aligned}
\|x_k - z\| \\
&\le \delta + \|z - P_{k-1}(x_{k-1} - \alpha_{k-1}\xi_{k-1})\| \\
&\le \delta + \|z - x_{k-1}\| + \alpha_{k-1}\|\xi_{k-1}\| \\
&\le 2\delta + \beta_0\|\xi_{k-1}\|.
\end{aligned}
\tag{2.437}
$$

Lemma 2.19, (2.401), (2.414), and (2.415) imply that

$$\partial_\delta f(x_{k-1}) \subset B_X(0, L_0 + \delta) \subset B_X(0, L_0 + 1). \tag{2.438}$$

In view of (2.412) and (2.438),

$$\|\xi_{k-1}\| \le L_0 + 2. \tag{2.439}$$

By (2.404), (2.408), (2.435), (2.437), and (2.439),

$$\|x_k - z\| \le 2\delta + \beta_0(L_0 + 2), \tag{2.440}$$

$$d(x_k, C) \le 2\delta + \beta_0(L_0 + 2) < \bar{\epsilon}. \tag{2.441}$$

By (2.436) and (2.440),

$$
\begin{aligned}
\|x_k - x_{k-1}\| &\le \|x_k - z\| + \|z - x_{k+1}\| \\
&\le 3\delta + \beta_0(L_0 + 2).
\end{aligned}
\tag{2.442}
$$

In view of (2.401), (2.414), (2.415), and (2.442),

$$|f(x_{k-1}) - f(x_k)| \le L_0\|x_k - x_{k-1}\| \le 3L_0\delta + \beta_0 L_0(L_0 + 2). \tag{2.443}$$

In view of (2.404), (2.408), and (2.443),

$$
\begin{aligned}
f(x_k) &\le f(x_{k-1}) + 3L_0\delta + \beta_0 L_0(L_0 + 2) \\
&\le \inf(f, C) + \bar{\epsilon} + 8^{-1}\bar{\epsilon} + 8^{-1}\bar{\epsilon} < \inf(f, C) + 2\bar{\epsilon}.
\end{aligned}
\tag{2.444}
$$

It follows from (2.403), (2.441), and (2.444) that

$$d(x_k, C_{min}) \le \epsilon.$$

This inequality contradicts (2.429). The contradiction we have reached proves (2.434).

It follows from (2.401), (2.407), (2.408), (2.412)–(2.415), (2.432), and (2.434) that Lemma 2.22 holds with

$$P_C = P_k, \ K_0 = M + 2\bar{K} + 2, \ x = x_{k-1}, \ y = x_k, \ \xi = \xi_{k-1},$$

$$\alpha = \alpha_{k-1}, \ \epsilon = \bar{\epsilon}, \ \delta_f = \delta, \ \delta_C = \delta$$

, and combining with (2.404) and (2.411), this implies that

$$
\begin{aligned}
d(x_k, &C_{min})^2 \\
&\leq d(x_{k-1}, C_{min})^2 - 2^{-1}\alpha_{k-1}\bar{\epsilon} + 25\alpha_{k-1}^2(L_0 + \bar{L})^2 + \delta^2 \\
&\quad + 2\delta(M + 3\bar{K} + 2 + 5L_0 + 5\bar{L}) \\
&\leq d(x_{k-1}, C_{min})^2 - 4^{-1}\alpha_{k-1}\bar{\epsilon} + 2\delta(M + 3\bar{K} + 3 + 5L_0 + 5\bar{L}) \\
&\leq d(x_{k-1}, C_{min})^2 - 4^{-1}\beta_1\bar{\epsilon} + 2\delta(M + 3\bar{K} + 3 + 5L_0 + 5\bar{L}) \\
&\leq d(x_{k-1}, C_{min})^2 - 8^{-1}\beta_1\bar{\epsilon}
\end{aligned}
$$

and

$$d(x_k, C_{min}) \leq d(x_{k-1}, C_{min}) \leq \epsilon.$$

This contradicts (2.429). The contradiction we have reached proves that

$$d(x_i, C_{min}) \leq \epsilon$$

for all integers i satisfying $j \leq i \leq n$. Theorem 2.15 is proved.

2.18   Proof of Theorem 2.16

By (2.94), (2.95), (2.97), and (2.99), for each integer $i \geq 0$,

$$\|x_i\| \leq K_1, \tag{2.445}$$

$$\|\xi_i\| \leq L_1. \tag{2.446}$$

In view of (2.10), (2.96), (2.99), and (2.446), for each integer $i \geq 1$,

$$x_i \in C, \tag{2.447}$$

$$\|x_i - x_{i+1}\| = \|x_i - P_i(x_i - \alpha_i \xi_i)\|$$
$$\leq \|x_i - (x_i - \alpha_i \xi_i)\| \leq \alpha_i \|\xi_i\| \leq \alpha_i L_1. \tag{2.448}$$

By (2.448),

$$\sum_{i=1}^{\infty} \|x_i - x_{i+1}\| \leq L_1 \sum_{i=1}^{\infty} \alpha_i < \infty.$$

Thus $\{x_i\}_{i=1}^{\infty}$ is a Cauchy sequence, and there exists

$$x_* = \lim_{i \to \infty} x_i. \tag{2.449}$$

Assume that

$$x_* \notin C_{min}. \tag{2.450}$$

It follows from (2.447) and (2.449) that

$$x_* \in C. \tag{2.451}$$

In view of (2.450) and (2.451), there exists $\epsilon \in (0, 1)$ such that

$$f(x_*) > \inf(f, C) + \epsilon. \tag{2.452}$$

In view of (2.449), there exist a natural number n_0 such that for all integers $n \geq n_0$,

$$f(x_n) > \inf(f, C) + \epsilon, \tag{2.453}$$

$$\alpha_n \leq 100^{-1}(L_1 + \bar{L})^{-2}\epsilon. \tag{2.454}$$

Fix

$$z \in C_{min}. \tag{2.455}$$

Applying Lemma 2.21 with $K_0 = K_1$, $L_0 = L_1$, $x = x_t$, $y = x_{t+1}$, and $v = \xi_t$ and with arbitrary sufficiently small $\delta_f, \delta_C > 0$, we obtain that for all integers $t \geq n_0$,

$$\|x_{t+1} - z\|^2 \leq \|x_t - z\|^2 - 2^{-1}\alpha_t \epsilon + 25\alpha_t^2(L_1 + \bar{L})^2$$
$$\leq \|x_t - z\|^2 - 4^{-1}\alpha_t \epsilon.$$

Theorem 2.16 is proved.

2.19 Proof of Theorem 2.17

Fix

$$\bar{x} \in C_{min}. \tag{2.456}$$

We show that for all $i = 1, 2, \ldots,,$

$$d(x_i, C_{min}) \leq 1 + K_1 + \bar{K}.$$

We have

$$\|x_0\| \leq K_1. \tag{2.457}$$

By (2.7), (2.8), (2.10), (2.101), (2.104), (2.456), and (2.457),

$$\|x_1 - \bar{x}\| = \|\bar{x} - P_1(x_0 - \alpha_0 \xi_0)\| \leq \|\bar{x}_0 - x_0\| + \alpha_0 \|\xi_0\|. \tag{2.458}$$

In view of (2.100), (2.103), and (2.457),

$$\|\xi_0\| \leq L_1 + 1. \tag{2.459}$$

It follows from (2.7), (2.8), (2.102), and (2.457)–(2.459) that

$$\|x_1 - \bar{x}\| \leq K_1 + \bar{K} + 1. \tag{2.460}$$

Assume that $t \geq 1$ is an integer and

$$d(x_t, C_{min}) \leq K_1 + \bar{K} + 1. \tag{2.461}$$

In view of (2.456) and (2.461),

$$\|x_t\| \leq K_1 + 2\bar{K} + 1. \tag{2.462}$$

There are two cases:

$$f(x_t) \leq \inf(f, C) + 4; \tag{2.463}$$

$$f(x_t) > \inf(f, C) + 4. \tag{2.464}$$

Assume that (2.463) holds. In view of (2.7), (2.8), and (2.463),

$$\|x_t\| \leq \bar{K}. \tag{2.465}$$

In view of (2.465),

$$\|\bar{x} - x_t\| \le 2\bar{K}.$$ (2.466)

By (2.10), (2.101), (2.104), and (2.466),

$$\|x_{t+1} - \bar{x}\| = \|P_t(x_t - \alpha_t \xi_t) - \bar{x}\|$$
$$\le \|x_t - \bar{x}\| + \alpha_t \|\xi_t\|.$$ (2.467)

It follows from (2.100), (2.103), and (2.465) that

$$\|\xi_t\| \le L_1.$$ (2.468)

It follows from (2.102) and (2.466)–(2.468) that

$$\|x_{t+1} - \bar{x}\| \le 2\bar{K} + \alpha_t L_1 \le 2\bar{K} + 1.$$ (2.469)

Assume that (2.464) holds. In view of (2.7), (2.8), (2.102), (2.461), and (2.464), we apply Lemma 2.23 with

$$P_C = P_t, \quad K_0 = 3K_1 + 1, \quad L_0 = L_1, \quad \alpha = \alpha_t, \quad \xi = \xi_t, \quad x = x_t, \quad y = x_{t+1}$$

and with arbitrary sufficiently small positive δ_f, δ_C and obtain that

$$d(x_{t+1}, C_{min})^2 \le d(x_t, C_{min})^2 - 2\alpha_t + 25\alpha_t^2(L_1 + \bar{L})^2$$
$$\le d(x_t, C_{min})^2 - \alpha_t \le d(x_t, C_{min})^2$$

and

$$d(x_{t+1}, C_{min}) \le K_1 + \bar{K} + 1.$$

Together with (2.456) and (2.459), this implies that the inequality above holds in the both cases. Therefore by induction we showed that

$$d(x_t, C_{min}) \le K_1 + \bar{K} + 1 \text{ for all integers } t \ge 0.$$ (2.470)

In view of (2.7), (2.8), (2.456), and (2.470),

$$\|x_t\| \le K_1 + 2\bar{K} + 1 \text{ for all integers } t \ge 0.$$ (2.471)

It follows from (2.100), (2.103), and (2.471) that

$$\|\xi_t\| \le L_1 \text{ for all integers } t \ge 0.$$ (2.472)

By (2.10), (2.101), (2.104), and(2.472),

$$\|x_{t+1} - x_t\| = \|P_t(x_t - \alpha_t \xi_t) - x_t\| \leq \alpha_t \|\xi_t\| \leq \alpha_t L_1.$$

Therefore

$$\sum_{t=0}^{\infty} \|x_{t+1} - x_t\| \leq L_1 \sum_{t=0}^{\infty} \alpha_t < \infty. \tag{2.473}$$

This implies that there exists

$$x_* = \lim_{t \to \infty} x_t \in C. \tag{2.474}$$

Assume that

$$x_* \notin C_{min}.$$

By the relation above, there exists $\epsilon \in (0, 1)$ such that

$$f(x_*) > \inf(f, C) + \epsilon. \tag{2.475}$$

In view of (2.474) and (2.475), there exist a natural number n_0 such that for all integers $n \geq n_0$,

$$f(x_n) > \inf(f, C) + \epsilon, \tag{2.476}$$

$$\alpha_n \leq 100^{-1}(L_1 + \bar{L})^{-2}\epsilon. \tag{2.477}$$

Let $z \in C_{min}$. In view of (2.100), (2.103), (2.104), (2.427), (2.456), and (2.471), we apply Lemma 2.21 with $K_0 = 3K_1$, $L_0 = L_1$, $x = x_t$, $y = x_{t+1}$, and $\xi = \xi_t$ and with arbitrary sufficiently small $\delta_f, \delta_C > 0$ and obtain that for all integers $t \geq n_0$,

$$\|x_{t+1} - z\|^2 \leq \|x_t - z\|^2 - 2^{-1}\alpha_t \epsilon + 25\alpha_t^2(L_1 + \bar{L})^2$$

$$\leq \|x_t - z\|^2 - 4^{-1}\alpha_t \epsilon.$$

Theorem 2.17 is proved.

2.20 Proof of Theorem 2.18

By (2.10), (2.105), (2.106), and (2.108), for each integer $t \geq 0$,

$$\|x_t - x_{t+1}\| = \|x_t - P_t(x_t - \alpha_t \|\xi_t\|^{-1}\xi_t)\| \le \alpha_t, \tag{2.478}$$

$$\sum_{t=0}^{\infty} \|x_t - x_{t+1}\| \le \sum_{t=0}^{\infty} \alpha_t < \infty.$$

Thus $\{x_t\}_{t=0}^{\infty}$ is a Cauchy sequence, and there exists

$$x_* \in \lim_{t \to \infty} x_t \in C. \tag{2.479}$$

By (2.106) and (2.478), for all integers $t \ge 0$,

$$\|x_t\| \le \|x_0\| + \sum_{i=0}^{t} \alpha_i \le K_1 + \sum_{i=0}^{\infty} \alpha_i. \tag{2.480}$$

Assume that

$$x_* \notin C_{min}. \tag{2.481}$$

In view of (2.479) and (2.481), there exists $\epsilon > 0$ such that

$$f(x_*) > \inf(f, C) + \epsilon. \tag{2.482}$$

In view of (2.479) and (2.482), there exist a natural number n_0 such that for all integers $n \ge n_0$,

$$f(x_n) > \inf(f, C) + \epsilon, \tag{2.483}$$

$$\alpha_n \le 16^{-1} \bar{L}^{-1} \epsilon. \tag{2.484}$$

Fix

$$z \in C_{min} \tag{2.485}$$

and $t \ge n_0$ be an integer. In view of (2.107), (2.108), (2.480), and (2.483)–(2.485), we apply Lemma 2.24 with $\bar{x} = z$, $K_0 = K_1 + \sum_{i=0}^{\infty} \alpha_i$, $\alpha = \alpha_t$, $x = x_t$, $y = x_{t+1}$, and $\xi = \xi_t$ and with arbitrary sufficiently small $\delta_f, \delta_C > 0$ and obtain that

$$\|x_{t+1} - z\|^2 \le \|x_t - z\|^2 - (4\bar{L})^{-1}\alpha_t \epsilon + 2\alpha_t^2$$
$$\le \|x_t - z\|^2 - (4\bar{L})^{-1}\alpha_t \epsilon.$$

Theorem 2.18 is proved.

Chapter 3
Extensions

In this chapter we study the projected subgradient method for nonsmooth convex constrained optimization problems in a Hilbert space. For these problems, an objective function is defined on an open convex set and a set of admissible points is not necessarily convex. We generalize some results of Chapter 2 obtained in the case when an objective function is defined on the whole Hilbert space.

3.1 Optimization Problems on Bounded Sets

Let $(X, \langle \cdot, \cdot \rangle)$ be a Hilbert space with an inner product $\langle \cdot, \cdot \rangle$ which induces a complete norm $\| \cdot \|$. We use the notation and definitions introduced in Chapter 2.

Let C be a closed nonempty subset of the space X and U be an open convex subset of X such that

$$C \subset U. \tag{3.1}$$

Suppose that $L, M > 0$,

$$C \subset B_X(0, M), \tag{3.2}$$

and that a convex function $f : U \to R^1$ satisfies

$$|f(u_1) - f(u_2)| \leq L\|u_1 - u_2\| \text{ for all } u_1, u_2 \in U. \tag{3.3}$$

For each point $x \in U$ and each positive number ϵ, let

$$\partial f(x) = \{l \in X : f(y) - f(x) \geq \langle l, y - x \rangle \text{ for all } y \in U\} \tag{3.4}$$

© The Editor(s) (if applicable) and The Author(s), under exclusive license to
Springer Nature Switzerland AG 2020
A. J. Zaslavski, *The Projected Subgradient Algorithm in Convex Optimization*,
SpringerBriefs in Optimization, https://doi.org/10.1007/978-3-030-60300-7_3

and let

$$\partial_\epsilon f(x) = \{l \in X : \ f(y) - f(x) \geq \langle l, y - x \rangle - \epsilon \text{ for all } y \in U\}. \tag{3.5}$$

Denote by $\mathcal{M}$ the set of all mappings $P : X \to C$ such that

$$Pz = z \text{ for all } z \in C,$$

$$\|Px - z\| \leq \|x - z\| \text{ for all } x \in X \text{ and all } z \in C. \tag{3.6}$$

Define

$$\inf(f, C) = \inf\{f(z) : \ z \in C\}. \tag{3.7}$$

It is clear that $\inf(f, C)$ is finite.
Set

$$C_{min} = \{x \in C : \ f(x) = \inf(f, C)\}. \tag{3.8}$$

For all $P \in \mathcal{M}$ set $P^0 x = x, x \in X$. We assume that

$$C_{min} \neq \emptyset.$$

In view of (3.3), for each $x \in U$,

$$\partial f(x) \subset B_X(0, L). \tag{3.9}$$

Proposition 3.1 *Assume that* $\epsilon, r > 0$ *and that* $x \in U$,

$$B_X(x, r) \subset U. \tag{3.10}$$

Then

$$\partial_\epsilon f(x) \subset B_X(0, L + \epsilon r^{-1}).$$

Proof Let $\xi \in \partial_\epsilon f(x)$. By (3.5) and (3.10), for every $h \in B_X(0, 1)$,

$$Lr \geq Lr\|h\| \geq f(x + rh) - f(x) \geq \langle \xi, rh \rangle - \epsilon,$$

$$\langle \xi, h \rangle \leq L + \epsilon/r.$$

This implies that $\|\xi\| \leq L + \epsilon/r$. Proposition 3.1 is proved.

In this chapter we prove the following two results.

Theorem 3.2 *Assume that $\delta_f, \delta_C \in (0, 1]$, $T \geq 1$ is an integer, $\{\alpha_t\}_{t=0}^{T-1} \subset (0, 1]$,*

$$\{P_i\}_{i=0}^{T-1} \subset \mathcal{M}, \tag{3.11}$$

$$\{x_i\}_{i=0}^{T} \subset U, \ \{\xi_i\}_{i=0}^{T-1} \subset X,$$

$$\|x_0\| \leq M + 1, \tag{3.12}$$

and that for $i = 0, \ldots, T - 1$,

$$B_X(\xi_i, \delta_f) \cap \partial f(x_i) \neq \emptyset, \tag{3.13}$$

$$\|x_{i+1} - P_i(x_i - \alpha_i \xi_i)\| \leq \delta_C. \tag{3.14}$$

Then

$$\min\{f(x_t) : t = 0, \ldots, T - 1\} - \inf(f, C), \ f\left(\left(\sum_{i=0}^{T-1} \alpha_i\right)^{-1} \sum_{t=0}^{T-1} \alpha_t x_t\right) - \inf(f, C)$$

$$\leq \left(\sum_{j=0}^{T-1} \alpha_j\right)^{-1} \sum_{t=0}^{T-1} \alpha_t (f(x_t) - \inf(f, C))$$

$$\leq 2^{-1}(2M + 1)^2 \left(\sum_{t=0}^{T-1} \alpha_t\right)^{-1} + 2^{-1} L^2 \left(\sum_{t=0}^{T-1} \alpha_t^2\right) \left(\sum_{t=0}^{T-1} \alpha_t\right)^{-1}$$

$$+ T\delta_C \left(\sum_{t=0}^{T-1} \alpha_t\right)^{-1} (2M + L + 3) + \delta_f (2M + L + 2). \tag{3.15}$$

Theorem 3.3 *Assume that $r > 0$,*

$$B_X(z, 2r) \subset U \text{ for all } z \in C, \tag{3.16}$$

$\Delta > 0$, $\delta_f, \delta_C \in (0, 1]$, $\delta_C \leq r$, $T \geq 1$ *is an integer,* $\{\alpha_t\}_{t=0}^{T-1} \subset (0, 1]$,

$$\{P_i\}_{i=0}^{T-1} \subset \mathcal{M}, \tag{3.17}$$

$$\{x_i\}_{i=0}^{T} \subset U, \ \{\xi_i\}_{i=0}^{T-1} \subset X,$$

$$\|x_0\| \leq M + 1, \tag{3.18}$$

$$B_X(x_0, r) \subset U, \tag{3.19}$$

and that for i $= 0, \ldots, T - 1$,

$$B_X(\xi_i, \delta_f) \cap \partial_\Delta f(x_i) \neq \emptyset, \tag{3.20}$$

$$\|x_{i+1} - P_i(x_i - \alpha_i\xi_i)\| \leq \delta_C. \tag{3.21}$$

Then

$$\min\{f(x_t) : t = 0, \ldots, T - 1\} - \inf(f, C), \; f\left(\left(\sum_{i=0}^{T-1}\alpha_i\right)^{-1}\sum_{t=0}^{T-1}\alpha_t x_t\right) - \inf(f, C)$$

$$\leq \left(\sum_{j=0}^{T-1}\alpha_j\right)^{-1}\sum_{t=0}^{T-1}\alpha_t(f(x_t) - \inf(f, C))$$

$$\leq 2^{-1}(2M + 1)^2\left(\sum_{t=0}^{T-1}\alpha_t\right)^{-1} + \Delta + 2^{-1}(L + \Delta r^{-1})^2\left(\sum_{t=0}^{T-1}\alpha_t^2\right)\left(\sum_{t=0}^{T-1}\alpha_t\right)^{-1}$$

$$+ T\delta_C\left(\sum_{t=0}^{T-1}\alpha_t\right)^{-1}(2M + L + 3 + \Delta r^{-1}) + \delta_f(2M + L + 2 + \Delta r^{-1}). \tag{3.22}$$

Note that (3.15) is a particular case of (3.22) with $\Delta = 0$.

Theorems 3.2 and 3.3 are new. They are proved in Sections 3.4 and 3.5, respectively.

Let $T \geq 1$ be an integer and $A > 0$ be given. We are interested in an optimal choice of α_t, $t = 0, \ldots, T - 1$ satisfying $\sum_{t=0}^{T-1}\alpha_t = A$ which minimizes the right-hand side of (3.22). By Lemma 2.3 of [75], $\alpha_t = \alpha = T^{-1}A$, $t = 0, \ldots, T - 1$. In this case the right-hand side of (3.22) is

$$2^{-1}(2M + 1)^2 T^{-1}\alpha^{-1} + \Delta + 2^{-1}(L + \Delta r^{-1})^2\alpha$$

$$+ \delta_C\alpha^{-1}(2M + L + 3 + \Delta r^{-1}) + \delta_f(2M + L + 2 + \Delta r^{-1}).$$

Now we can make the best choice of the step-size $\alpha > 0$. Since T can be arbitrarily large, we need to minimize the function

$$\delta_C\alpha^{-1}(2M + L + 3 + \Delta r^{-1}) + 2^{-1}(L + \Delta r^{-1})^2\alpha, \; \alpha > 0$$

which has a minimizer

$$\alpha = (L + \Delta r^{-1})^{-1}(2\delta_C(2M + L + 3 + \Delta r^{-1}))^{1/2}.$$

With this choice of α the right-hand side of (3.22) is

$$2^{-1}(2M+1)^2 T^{-1}(L+\Delta r^{-1})(2\delta_C(2M+L+3+\Delta r^{-1}))^{1/2}+\Delta$$
$$+(L+\Delta r^{-1})(2^{-1}\delta_C(2M+L+3+\Delta r^{-1}))^{1/2}$$
$$+\delta_f(2M+L+2+\Delta r^{-1})$$
$$+2^{-1}(L+\Delta r^{-1})(2\delta_C(2M+L+3+\Delta r^{-1}))^{1/2}.$$

Now we should make the best choice of T. It is clear that T should be at the same order as δ_C^{-1}. In this case the right-hand side of (3.22) does not exceed $c_1\delta_C^{1/2}+\Delta+\delta_f(2M+L+2+\Delta r^{-1})$, where $c_1 > 0$ is a constant.

3.2   An Auxiliary Result for Theorem 3.2

Lemma 3.4 *Let*

$$P \in \mathcal{M}, \tag{3.23}$$

$\alpha \in (0,1]$, $\delta_f, \delta_C \in (0,1]$

$$x \in U \cap B_X(0, M+1), \tag{3.24}$$

let $\xi \in X$ satisfy

$$B_X(\xi, \delta_f) \cap \partial f(x) \neq \emptyset, \tag{3.25}$$

and let

$$y \in B_X(P(x - \alpha\xi), \delta_C). \tag{3.26}$$

Then for each $z \in C$,

$$2\alpha(f(x) - f(z)) \leq \|x - z\|^2 - \|y - z\|^2$$
$$+ \alpha^2 L^2 + 2\delta_C(2M + L + 3) + 2\alpha\delta_f(2M + L + 2).$$

Proof Let

$$z \in C. \tag{3.27}$$

By (3.25), there exists

$$v \in \partial f(x) \tag{3.28}$$

such that

$$\|\xi - v\| \le \delta_f. \tag{3.29}$$

In view of (3.9) and (3.28),

$$\|v\| \le L. \tag{3.30}$$

By (3.29) and (3.30),

$$\|\xi\| \le L + 1. \tag{3.31}$$

It follows from (3.2), (3.24), and (3.27)–(3.30) that

$$
\begin{aligned}
\|x - &\alpha\xi - z\|^2 \\
&= \|x - \alpha v + (\alpha v - \alpha\xi) - z\|^2 \\
&\le \|x - \alpha v - z\|^2 + \alpha^2\|v - \xi\|^2 + 2\alpha\langle v - \xi, x - \alpha v - z\rangle \\
&\le \|x - \alpha v - z\|^2 + \alpha^2\delta_f^2 + 2\alpha\delta_f\|x - \alpha v - z\| \\
&\le \|x - \alpha v - z\|^2 + \alpha^2\delta_f^2 + 2\alpha\delta_f(2M + L + 1) \\
&\le \|x - z\|^2 - 2\alpha\langle x - z, v\rangle + \alpha^2\|v\|^2 \\
&\quad + \alpha^2\delta_f^2 + 2\alpha\delta_f(2M + 1 + L) \\
&\le \|x - z\|^2 - 2\alpha\langle x - z, v\rangle + \alpha^2 L^2 \\
&\quad + \alpha^2\delta_f^2 + 2\alpha\delta_f(2M + 1 + L). \tag{3.32}
\end{aligned}
$$

In view of (3.28),

$$\langle v, z - x\rangle \le f(z) - f(x). \tag{3.33}$$

By (3.32) and (3.33),

$$
\begin{aligned}
\|x - \alpha\xi - z\|^2 &\le \|x - z\|^2 \\
&\quad + 2\alpha(f(z) - f(x)) + \alpha^2 L^2 + 2\alpha\delta_f(2M + 2 + L). \tag{3.34}
\end{aligned}
$$

It follows from (3.2), (3.24), (3.27), and (3.31) that

$$\|x - \alpha\xi - z\| \le 2M + 2 + L. \tag{3.35}$$

By (3.26), (3.27), (3.34), and (3.35),

$$
\begin{aligned}
\|y - z\|^2 &= \|y - P(x - \alpha\xi) + P(x - \alpha\xi) - z\|^2 \\
&\leq \|y - P(x - \alpha\xi)\|^2 + 2\|y - P(x - \alpha\xi)\|\|P(x - \alpha\xi) - z\| \\
&\quad + \|P(x - \alpha\xi) - z\|^2 \\
&\leq \delta_C^2 + 2\delta_C\|x - \alpha\xi - z\| + \|x - \alpha\xi - z\|^2 \\
&\leq \delta_C^2 + 2\delta_C(2M + L + 2) + \|x - z\|^2 \\
&\quad + 2\alpha(f(z) - f(x)) + \alpha^2 L^2 + 2\alpha\delta_f(2M + 2 + L).
\end{aligned}
$$

This implies that

$$
\begin{aligned}
2\alpha(f(x) - f(z)) &\leq \|x - z\|^2 - \|y - z\|^2 \\
&\quad + \alpha^2 L^2 + 2\delta_C(2M + L + 3) + 2\alpha\delta_f(2M + L + 2).
\end{aligned}
$$

Lemma 3.4 is proved.

3.3   An Auxiliary Result for Theorem 3.3

Lemma 3.5 *Let*

$$
P \in \mathcal{M},
$$

$\alpha \in (0, 1]$, $\delta_f, \delta_C \in (0, 1]$, $r, \Delta > 0$,

$$
x \in U \cap B_X(0, M + 1), \tag{3.36}
$$

$$
B_X(x, r) \subset U, \tag{3.37}
$$

$\xi \in X$ *satisfy*

$$
B_X(\xi, \delta_f) \cap \partial_\Delta f(x) \neq \emptyset, \tag{3.38}
$$

and let

$$
y \in B_X(P(x - \alpha\xi), \delta_C). \tag{3.39}
$$

Then for each $z \in C$,

$$2\alpha(f(x) - f(z)) \le \|x - z\|^2 - \|y - z\|^2$$
$$+ 2\alpha\Delta + \delta_C^2 + +2\delta_C(2M + L + 2 + \Delta r^{-1})$$
$$+ \alpha^2(L + \Delta r^{-1})^2 + 2\alpha\delta_f(2M + L + 2 + \Delta r^{-1}).$$

Proof Let

$$z \in C. \tag{3.40}$$

By (3.38), there exists

$$v \in \partial_\Delta f(x) \tag{3.41}$$

such that

$$\|\xi - v\| \le \delta_f. \tag{3.42}$$

Proposition 3.1, (3.36), and (3.37) imply that

$$\partial_\Delta f(x) \subset B_X(0, L + \Delta r^{-1}). \tag{3.43}$$

In view of (3.41) and (3.43),

$$\|v\| \le L + \Delta r^{-1}. \tag{3.44}$$

By (3.42) and (3.44),

$$\|\xi\| \le L + \Delta r^{-1} + 1. \tag{3.45}$$

It follows from (3.2), (3.36), (3.40), (3.42), and (3.44) that

$$\|x - \alpha\xi - z\|^2$$
$$= \|x - \alpha v + (\alpha v - \alpha\xi) - z\|^2$$
$$\le \|x - \alpha v - z\|^2 + \alpha^2\|v - \xi\|^2 + 2\alpha\langle v - \xi, x - \alpha v - z\rangle$$
$$\le \|x - \alpha v - z\|^2 + \alpha^2\delta_f^2 + 2\alpha\delta_f\|x - \alpha v - z\|$$
$$\le \|x - \alpha v - z\|^2 + \alpha^2\delta_f^2 + 2\alpha\delta_f(2M + L + 1 + \Delta r^{-1})$$
$$\le \|x - z\|^2 - 2\alpha\langle x - z, v\rangle + \alpha^2\|v\|^2$$
$$+ \alpha^2\delta_f^2 + 2\alpha\delta_f(2M + 1 + L + \Delta r^{-1})$$

$$\leq \|x - z\|^2 - 2\alpha \langle x - z, v \rangle + \alpha^2 (L + \Delta r^{-1})^2$$

$$+ \alpha^2 \delta_f^2 + 2\alpha \delta_f (2M + 1 + L + \Delta r^{-1}). \tag{3.46}$$

In view of (3.41),

$$\langle v, z - x \rangle \leq f(z) - f(x) + \Delta. \tag{3.47}$$

By (3.46) and (3.47),

$$\|x - \alpha\xi - z\|^2 \leq \|x - z\|^2$$

$$+ 2\alpha(f(z) - f(x)) + 2\alpha\Delta + \alpha^2 (L + \Delta r^{-1})^2$$

$$+ 2\alpha\delta_f (2M + 2 + L + \Delta r^{-1}). \tag{3.48}$$

It follows from (3.27), (3.36), (3.40), and (3.45) that

$$\|x - \alpha\xi - z\| \leq 2M + 2 + L + \Delta r^{-1}. \tag{3.49}$$

By (3.6), (3.39), (3.40), (3.48), and (3.49),

$$\|y - z\|^2 = \|y - P(x - \alpha\xi) + P(x - \alpha\xi) - z\|^2$$

$$\leq \|y - P(x - \alpha\xi)\|^2 + 2\|y - P(x - \alpha\xi)\| \|P(x - \alpha\xi) - z\|$$

$$+ \|P(x - \alpha\xi) - z\|^2$$

$$\leq \delta_C^2 + 2\delta_C \|x - \alpha\xi - z\| + \|x - \alpha\xi - z\|^2$$

$$\leq \delta_C^2 + 2\delta_C (2M + L + 2 + \Delta r^{-1}) + \|x - z\|^2$$

$$+ 2\alpha(f(z) - f(x)) + 2\alpha\Delta + \alpha^2 (L + \Delta r^{-1})^2$$

$$+ 2\alpha\delta_f (2M + 2 + L + \Delta r^{-1}).$$

This implies that

$$2\alpha(f(x) - f(z)) \leq \|x - z\|^2 - \|y - z\|^2 + 2\alpha\Delta + \delta_C^2$$

$$+ \alpha^2 (L + \Delta r^{-1})^2 + 2\delta_C (2M + L + 2 + \Delta r^{-1})$$

$$+ 2\alpha\delta_f (2M + L + 2 + \Delta r^{-1}).$$

Lemma 3.5 is proved.

3.4   Proof of Theorem 3.2

Fix

$$\bar{x} \in C_{min}.$$

For every $t = 0, \ldots, T - 1$ we apply Lemma 3.4 with $P = P_t$, $x = x_t$, $y = x_{t+1}$, $\xi = \xi_t$, $\alpha = \alpha_t$ and obtain that

$$2\alpha_t(f(x_t) - f(\bar{x})) \le \|x_t - \bar{x}\|^2 - \|x_{t+1} - \bar{x}\|^2$$
$$+ \alpha_t^2 L^2 + 2\delta_C(2M + L + 3) + 2\alpha_t\delta_f(2M + L + 2).$$

Together with (3.2) and (3.12), this implies that

$$\sum_{t=0}^{T-1} \alpha_t(f(x_t) - \inf(f, C)) \le 2^{-1} \sum_{t=0}^{T-1} (\|x_t - \bar{x}\|^2 - \|x_{t+1} - \bar{x}\|^2)$$

$$+ 2^{-1} \sum_{t=0}^{T-1} \alpha_t^2 L^2 + T\delta_C(2M + L + 3) + \delta_f(2M + L + 2) \sum_{t=0}^{T-1} \alpha_t$$

$$\le 2^{-1}(2M + 1)^2 + 2^{-1} \sum_{t=0}^{T-1} \alpha_t^2 L^2 + T\delta_C(2M + L + 3) + \delta_f(2M + L + 2).$$

This implies that

$$\min\{f(x_t) : t = 0, \ldots, T - 1\} - \inf(f, C), \ f\left(\left(\sum_{i=0}^{T-1} \alpha_i\right)^{-1} \sum_{t=0}^{T-1} \alpha_t x_t\right) - \inf(f, C)$$

$$\le \left(\sum_{j=0}^{T-1} \alpha_j\right)^{-1} \sum_{t=0}^{T-1} \alpha_t(f(x_t) - \inf(f, C))$$

$$\le 2^{-1}(2M + 1)^2 \left(\sum_{t=0}^{T-1} \alpha_t\right)^{-1} + 2^{-1}L^2 \left(\sum_{t=0}^{T-1} \alpha_t^2\right)\left(\sum_{t=0}^{n} \alpha_t\right)^{-1}$$

$$+ T\delta_C \left(\sum_{t=0}^{T-1} \alpha_t\right)^{-1}(2M + L + 3) + \delta_f(2M + L + 2).$$

Theorem 3.2 is proved.

3.5   Proof of Theorem 3.3

Fix

$$\bar{x} \in C_{min}.$$

By (3.2), (3.17), and (3.21), for all $t = 0, \ldots, T$,

$$\|x_t\| \le M + 1.$$

In view of (3.19),

$$B_X(x_0, r) \subset U.$$

It follows from (3.16), (3.17), and (3.21) that for all integers $t = 0, \ldots, T - 1$,

$$B_X(x_{t+1}, r) \subset B_X(P_t(x_t - \alpha_t \xi_t), r + \delta_C)$$
$$\subset B_X(P_t(x_t - \alpha_t \xi_t), 2r) \subset U.$$

For every $t = 0, \ldots, T - 1$ in view of the relation above, we apply Lemma 3.5 with $P = P_t$, $x = x_t$, $y = x_{t+1}$, $\alpha = \alpha_t$, $\xi = \xi_t$ and obtain that

$$2\alpha(f(x_t) - \inf(f, C)) \le \|x_t - \bar{x}\|^2 - \|x_{t+1} - \bar{x}\|^2$$
$$+ 2\alpha_t \Delta + \delta_C^2 + \alpha_t^2 (L + \Delta r^{-1})^2$$
$$+ 2\delta_C (2M + L + 2 + \Delta r^{-1}) + 2\alpha_t \delta_f (2M + L + 2 + \Delta r^{-1}).$$

The relation above implies that

$$\sum_{t=0}^{T-1} \alpha_t (f(x_t) - \inf(f, C)) \le 2^{-1} \sum_{t=0}^{T-1} (\|x_t - \bar{x}\|^2 - \|x_{t+1} - \bar{x}\|^2)$$

$$+ \sum_{t=0}^{T-1} \alpha_t \Delta + 2^{-1} \sum_{t=0}^{T-1} \alpha_t^2 (L + \Delta r^{-1})^2 + T\delta_C (2M + L + 3 + \Delta r^{-1})$$

$$+ \delta_f (2M + L + 2 + \Delta r^{-1}) \sum_{t=0}^{T-1} \alpha_t$$

$$\le 2^{-1} \|x_0 - \bar{x}\|^2 + \sum_{t=0}^{T-1} \alpha_t \Delta + 2^{-1} \sum_{t=0}^{T-1} \alpha_t^2 (L + \Delta r^{-1})^2$$

$$+ T\delta_C(2M + L + 3 + \Delta r^{-1})$$

$$+ \delta_f(2M + L + 2 + \Delta r^{-1}) \sum_{t=0}^{T-1} \alpha_t.$$

Together with (3.2) and (3.18) this implies that

$$\min\{f(x_t) : t = 0, \ldots, T - 1\} - \inf(f, C), \ f\left(\left(\sum_{i=0}^{T-1} \alpha_i\right)^{-1} \sum_{t=0}^{T-1} \alpha_t x_t\right) - \inf(f, C)$$

$$\leq \left(\sum_{j=0}^{T-1} \alpha_j\right)^{-1} \sum_{t=0}^{T-1} \alpha_t(f(x_t) - \inf(f, C))$$

$$\leq 2^{-1}(2M + 1)^2 \left(\sum_{t=0}^{T-1} \alpha_t\right)^{-1} + \Delta + 2^{-1}(L + \Delta r^{-1})^2 \left(\sum_{t=0}^{T-1} \alpha_t^2\right)\left(\sum_{t=0}^{n} \alpha_t\right)^{-1}$$

$$+ T\delta_C \left(\sum_{t=0}^{T-1} \alpha_t\right)^{-1} (2M + L + 3 + \Delta r^{-1}) + \delta_f(2M + L + 2 + \Delta r^{-1}).$$

Theorem 3.3 is proved.

3.6   Optimization on Unbounded Sets

Let $(X, \langle \cdot, \cdot \rangle)$ be a Hilbert space with an inner product $\langle \cdot, \cdot \rangle$ which induces a complete norm $\| \cdot \|$.

Let C be a closed nonempty subset of the space X, U be an open convex subset of X such that

$$C \subset U,$$

and $f : U \to R^1$ be a convex function which is Lipschitz on all bounded subsets of U.

For each point $x \in U$ and each positive number ϵ, let

$$\partial f(x) = \{l \in X : \ f(y) - f(x) \geq \langle l, y - x \rangle \text{ for all } y \in U\}$$

and let

$$\partial_\epsilon f(x) = \{l \in X : \ f(y) - f(x) \geq \langle l, y - x \rangle - \epsilon \text{ for all } y \in U\}.$$

Assume that

$$\lim_{x \in U, \|x\| \to \infty} f(x) = \infty. \tag{3.50}$$

It means that for each $M_0 > 0$ there exists $M_1 > 0$ such that if a point $x \in U$ satisfies the inequality $\|x\| \geq M_1$, then $f(x) > M_0$.

Define

$$\inf(f, C) = \inf\{f(z) : z \in C\}.$$

Since the function f is Lipschitz on all bounded subsets of the space X, it follows from (3.50) that $\inf(f, C)$ is finite.

Set

$$C_{min} = \{x \in C : f(x) = \inf(f, C)\}. \tag{3.51}$$

It is well-known that if the set C is convex, then the set C_{min} is nonempty. Clearly, the set $C_{min} \neq \emptyset$ if the space X is finite-dimensional.

We assume that

$$C_{min} \neq \emptyset.$$

It is clear that C_{min} is a closed subset of C. Fix

$$\theta_0 \in C. \tag{3.52}$$

Set

$$U_0 = \{x \in U : f(x) \leq f(\theta_0) + 4\}. \tag{3.53}$$

In view of (3.50) there exists a number $\bar{K} > 1$ such that

$$U_0 \subset B_X(0, \bar{K}). \tag{3.54}$$

Since the function f is Lipschitz on all bounded subsets of U, there exists a number $\bar{L} > 1$ such that

$$|f(z_1) - f(z_2)| \leq \bar{L}\|z_1 - z_2\| \text{ for all } z_1, z_2 \in U \cap B_X(0, \bar{K} + 4). \tag{3.55}$$

Denote by $\mathcal{M}_C$ the set of all mappings $P : X \to C$ such that

$$Pz = z \text{ for all } z \in C, \tag{3.56}$$

$$\|Pz - x\| \leq \|z - x\| \text{ for all } x \in C \text{ and all } z \in X. \tag{3.57}$$

We prove the following two theorems.

Theorem 3.6 *Assume that*

$$K_1 \geq \bar{K} + 4, \ L_1 \geq \bar{L}, \tag{3.58}$$

$\delta_f, \delta_C \in (0, 1]$,

$$|f(z_1) - f(z_2)| \leq L_1 \|z_1 - z_2\| \text{ for all } z_1, z_2 \in B_X(0, 3K_1 + 2) \cap U, \tag{3.59}$$

$$\alpha \in (0, (1 + \bar{L})^{-2}) \tag{3.60}$$

and that

$$\delta_f(\bar{K} + 3K_1 + 2 + L_1) \leq \alpha,$$

$$\delta_C(\bar{K} + 3K_1 + L_1 + 3) \leq \alpha. \tag{3.61}$$

Let $T \geq 2$ be an integer

$$\{P_t\}_{t=0}^{T-1} \subset \mathcal{M}_C, \tag{3.62}$$

$$\{x_t\}_{t=0}^T \subset U, \ \{\xi_t\}_{t=0}^{T-1} \subset X, \tag{3.63}$$

$$\|x_0\| \leq K_1, \tag{3.64}$$

$$B_X(x_0, \delta_C) \cap C \neq \emptyset \tag{3.65}$$

and that for $t = 0, \ldots, T - 1$,

$$B_X(\xi_t, \delta_f) \cap \partial f(x_t) \neq \emptyset, \tag{3.66}$$

$$\|x_{t+1} - P_t(x_t - \alpha \xi_t)\| \leq \delta_C. \tag{3.67}$$

Then

$$\|x_t\| \leq 2\bar{K} + K_1, \ t = 0, \ldots, T$$

and

$$\min\{f(x_t) : t = 0, \ldots, T - 1\} - \inf(f, C), \ f\left(T^{-1} \sum_{i=0}^{T-1} x_i\right) - \inf(f, C)$$

$$\leq T^{-1} \sum_{i=0}^{T-1} f(x_i) - \inf(f, C)$$

$$\leq (2T\alpha)^{-1}(K_1 + \bar{K})^2 + L_1^2\alpha + \alpha^{-1}\delta_C(\bar{K} + 3K_1 + L_1 + 3)$$

$$+ \delta_f(3K_1 + \bar{K} + L_1 + 2).$$

Theorem 3.7 *Assume that*

$$K_1 \geq \bar{K} + 4, \ L_1 \geq \bar{L}, \ r_0 \in (0, 1],$$

$$B_X(z, r_0) \subset U, \ z \in C, \tag{3.68}$$

$$|f(z_1) - f(z_2)| \leq L_1\|z_1 - z_2\| \text{ for all } z_1, z_2 \in B_X(0, 3K_1 + 1) \cap U, \tag{3.69}$$

$$\Delta \in (0, r_0], \ \delta_f, \delta_C \in (0, 2^{-1}r_0], \ \alpha \in (0, (\bar{L} + 3)^{-2}], \tag{3.70}$$

and that

$$\delta_f(3\bar{K} + K_1 + 4 + L_1) \leq \alpha, \tag{3.71}$$

$$\delta_C(3\bar{K} + K_1 + L_1 + 2) \leq \alpha. \tag{3.72}$$

Let $T \geq 2$ be an integer

$$\{P_t\}_{t=0}^{T-1} \subset \mathcal{M}_C, \tag{3.73}$$

$$\{x_t\}_{t=0}^{T} \subset U, \ \{\xi_t\}_{t=0}^{T-1} \subset X, \tag{3.74}$$

$$\|x_0\| \leq K_1, \tag{3.75}$$

$$B_X(x_0, \delta_C) \cap C \neq \emptyset \tag{3.76}$$

and that for $i = 1, \ldots, T - 1$,

$$B_X(\xi_t, \delta_f) \cap \partial_\Delta f(x_t) \neq \emptyset, \tag{3.77}$$

$$\|x_{t+1} - P_t(x_t - \alpha\xi_t)\| \leq \delta_C. \tag{3.78}$$

Then

$$\|x_t\| \leq 2\bar{K} + K_1, \ t = 0, \ldots, T$$

and

$$\min\{f(x_t) : t = 0, \ldots, T - 1\} - \inf(f, C), \quad f\left(T^{-1} \sum_{i=0}^{T-1} x_i\right) - \inf(f, C)$$

$$\leq T^{-1} \sum_{i=0}^{T-1} f(x_i) - \inf(f, C)$$

$$\leq (2T\alpha)^{-1}(K_1 + \bar{K})^2 + (L_1 + 2)^2\alpha$$

$$+ \alpha^{-1}\delta_C(3\bar{K} + K_1 + L_1 + 2) + \Delta + \delta_f(K_1 + 3\bar{K} + L_1 + 4).$$

3.7   Auxiliary Results

Lemma 3.8 *Let $K_0, L_0 > 0$,*

$$|f(z_1) - f(z_2)| \leq L_0\|z_1 - z_2\| \text{ for all } z_1, z_2 \in B_X(0, K_0 + 1) \cap U, \quad (3.79)$$

$$x \in B_X(0, K_0) \cap U, \quad v \in \partial f(x). \quad (3.80)$$

Then

$$\|v\| \leq L_0.$$

Proof In view of (3.80), for all $u \in U$,

$$f(u) - f(x) \geq \langle v, u - x \rangle. \quad (3.81)$$

There exists $r \in (0, 1)$ such that

$$B_X(x, r) \subset U. \quad (3.82)$$

By (3.80) and (3.82),

$$B_X(x, r) \subset B_X(0, K_0 + 1). \quad (3.83)$$

It follows from (3.70), (3.80), (3.82), and (3.83) that for all $h \in B_X(0, 1)$,

$$x + rh \in U \cap B_X(0, K_0 + 1),$$

$$\langle v, rh \rangle \leq f(x + rh) - f(x) \leq L_0 r\|h\| \leq L_0 r,$$

$$\langle v, h \rangle \leq L_0.$$

Therefore $\|v\| \leq L_0$. Lemma 3.8 is proved.

Lemma 3.9 *Let* $K_0, L_0 > 0$, $r \in (0, 1]$, $\Delta > 0$,

$$|f(z_1) - f(z_2)| \leq L_0 \|z_1 - z_2\| \text{ for all } z_1, z_2 \in B_X(0, K_0 + 1) \cap U, \quad (3.84)$$

$$x \in B_X(0, K_0) \cap U, \quad (3.85)$$

$$B_X(x, r) \subset U. \quad (3.86)$$

Then

$$\partial_\Delta f(x) \subset B_X(0, L_0 + \Delta r^{-1}).$$

Proof Let

$$\xi \in \partial_\Delta f(x). \quad (3.87)$$

By (3.85) and (3.86),

$$B_X(x, r) \subset U \cap B_X(0, K_0 + 1). \quad (3.88)$$

In view of (3.88), for each $h \in B_X(0, 1)$,

$$x + rh \in U \cap B_X(0, K_0 + 1).$$

Together with (3.84) and (3.87) this implies that

$$L_0 r \geq L_0 r \|h\| \geq f(x + rh) - f(x) \geq \langle \xi, rh \rangle - \Delta,$$

$$L_0 + \Delta r^{-1} \geq \langle \xi, h \rangle.$$

This implies that

$$\|\xi\| \leq L_0 + \Delta r^{-1}.$$

Lemma 3.9 is proved.

Lemma 3.10 *Let* $P \in \mathcal{M}_c$, $K_0 \geq \bar{K}$, $L_0 > 0$,

$$|f(z_1) - f(z_2)| \leq L_0 \|z_1 - z_2\| \text{ for all } z_1, z_2 \in B_X(0, K_0 + 2) \cap U, \quad (3.89)$$

$\alpha \in (0, 1]$, $\delta_f, \delta_C \in (0, 1]$,

$$x \in U \cap B_X(0, K_0 + 1), \quad (3.90)$$

$\xi \in X$ *satisfy*

$$B_X(\xi, \delta_f) \cap \partial f(x) \neq \emptyset, \tag{3.91}$$

and let

$$y \in B_X(P(x - \alpha\xi), \delta_C) \cap U. \tag{3.92}$$

Then for each $z \in C$ satisfying $f(z) \leq f(\theta_0) + 4$,

$$2\alpha(f(x) - f(z)) \leq \|x - z\|^2 - \|y - z\|^2$$
$$+ 2\delta_C(K_0 + \bar{K} + L_0 + 3) + \alpha^2 L_0^2 + 2\alpha\delta_f(K_0 + \bar{K} + L_0 + 2).$$

Proof Let

$$z \in C \tag{3.93}$$

satisfy

$$f(z) \leq f(\theta_0) + 4. \tag{3.94}$$

In view of (3.52), (3.53), and (3.94),

$$\|z\| \leq \bar{K}. \tag{3.95}$$

By (3.91), there exists

$$v \in \partial f(x) \tag{3.96}$$

such that

$$\|\xi - v\| \leq \delta_f. \tag{3.97}$$

In view of (3.89), (3.90), and (3.96),

$$\|v\| \leq L_0. \tag{3.98}$$

By (3.97) and (3.98),

$$\|\xi\| \leq L_0. \tag{3.99}$$

It follows from (3.90), (3.95), (3.97), and (3.98) that

$$\|x - \alpha\xi - z\|^2$$

$$= \|x - \alpha v + (\alpha v - \alpha\xi) - z\|^2$$

$$\leq \|x - \alpha v - z\|^2 + \alpha^2\|v - \xi\|^2 + 2\alpha\langle v - \xi, x - \alpha v - z\rangle$$

$$\leq \|x - \alpha v - z\|^2 + \alpha^2\delta_f^2 + 2\alpha\delta_f\|x - \alpha v - z\|$$

$$\leq \|x - \alpha v - z\|^2 + \alpha^2\delta_f^2 + 2\alpha\delta_f(K_0 + \bar{K} + L_0 + 1)$$

$$\leq \|x - z\|^2 - 2\alpha\langle x - z, v\rangle + \alpha^2\|v\|^2$$

$$+ \alpha^2\delta_f^2 + 2\alpha\delta_f(K_0 + \bar{K} + L_0 + 1)$$

$$\leq \|x - z\|^2 - 2\alpha\langle x - z, v\rangle + \alpha^2 L_0^2$$

$$+ \alpha^2\delta_f^2 + 2\alpha\delta_f(K_0 + \bar{K} + L_0 + 1). \tag{3.100}$$

In view of (3.46),

$$\langle v, z - x\rangle \leq f(z) - f(x). \tag{3.101}$$

By (3.100) and (3.101),

$$\|x - \alpha\xi - z\|^2 \leq \|x - z\|^2$$

$$+ 2\alpha(f(z) - f(x)) + \alpha^2 L_0^2 + \alpha^2\delta_f^2 + 2\alpha\delta_f(K_0 + \bar{K} + L_0 + 1). \tag{3.102}$$

It follows from (3.90), (3.95), and (3.99) that

$$\|x - \alpha\xi - z\| \leq K_0 + 2 + \bar{K} + L_0. \tag{3.103}$$

By (3.56), (3.57), (3.92), and (3.103),

$$\|y - z\|^2 = \|y - P(x - \alpha\xi) + P(x - \alpha\xi) - z\|^2$$

$$\leq \|y - P(x - \alpha\xi)\|^2 + 2\|y - P(x - \alpha\xi)\|\|P(x - \alpha\xi) - z\|$$

$$+ \|P(x - \alpha\xi) - z\|^2$$

$$\leq \delta_C^2 + 2\delta_C(K_0 + \bar{K} + L_0 + 2) + \|x - \alpha\xi - z\|^2$$

$$\leq \delta_C^2 + 2\delta_C(K_0 + \bar{K} + L_0 + 2) + \|x - z\|^2$$

$$+ 2\alpha(f(z) - f(x)) + \alpha^2 L_0^2 + \alpha^2\delta_f^2 + 2\alpha\delta_f(K_0 + \bar{K} + L_0 + 1).$$

This implies that

$$2\alpha(f(x) - f(z)) \le \|x - z\|^2 - \|y - z\|^2 + 2\delta_C(K_0 + \bar{K} + L_0 + 3)$$
$$+ \alpha^2 L_0^2 + 2\alpha\delta_f(K_0 + \bar{K} + L_0 + 2).$$

Lemma 3.10 is proved.

Lemma 3.11 *Let* $P \in \mathcal{M}_c$, $K_0 \ge \bar{K}$, $L_0 > 0$,

$$|f(z_1) - f(z_2)| \le L_0\|z_1 - z_2\| \text{ for all } z_1, z_2 \in B_X(0, K_0 + 2) \cap U, \quad (3.104)$$

$\alpha \in (0, 1]$, $\delta_f, \delta_C \in (0, 1]$, $\Delta > 0$, $r \in (0, 1]$,

$$x \in U \cap B_X(0, K_0 + 1), \quad (3.105)$$

$$B_X(x, r) \subset U, \quad (3.106)$$

$\xi \in X$ *satisfy*

$$B_X(\xi, \delta_f) \cap \partial_\Delta f(x) \ne \emptyset, \quad (3.107)$$

and let

$$y \in B_X(P(x - \alpha\xi), \delta_C) \cap U. \quad (3.108)$$

Then for each $z \in C$ *satisfying* $f(z) \le f(\theta_0) + 4$,

$$2\alpha(f(x) - f(z)) \le \|x - z\|^2 - \|y - z\|^2$$
$$+ 2\delta_C(K_0 + \bar{K} + L_0 + 3 + \Delta r^{-1}) + 2\alpha\Delta + \alpha^2(L_0 + \Delta r^{-1})^2$$
$$+ 2\alpha\delta_f(K_0 + \bar{K} + L_0 + 2 + \Delta r^{-1}).$$

Proof Let

$$z \in C \quad (3.109)$$

satisfy

$$f(z) \le f(\theta_0) + 4. \quad (3.110)$$

In view of (3.53), (3.54), (3.109), and (3.110),

$$\|z\| \le \bar{K}. \quad (3.111)$$

By (3.107), there exists

$$v \in \partial_\Delta f(x) \tag{3.112}$$

such that

$$\|\xi - v\| \leq \delta_f. \tag{3.113}$$

Lemma 3.9 and (3.104)–(3.106) imply that

$$\partial_\Delta f(x) \subset B_X(0, L_0 + \Delta r^{-1}). \tag{3.114}$$

In view of (3.112) and (3.114),

$$\|v\| \leq L_0 + \Delta r^{-1}. \tag{3.115}$$

By (3.113) and (3.115),

$$\|\xi\| \leq L_0 + \Delta r^{-1} + 1. \tag{3.116}$$

It follows from (3.105), (3.111), (3.113), and (3.115) that

$$
\begin{aligned}
\|x &- \alpha\xi - z\|^2 \\
&= \|x - \alpha v + (\alpha v - \alpha\xi) - z\|^2 \\
&\leq \|x - \alpha v - z\|^2 + \alpha^2 \|v - \xi\|^2 + 2\alpha \langle v - \xi, x - \alpha v - z \rangle \\
&\leq \|x - \alpha v - z\|^2 + \alpha^2 \delta_f^2 + 2\alpha \delta_f \|x - \alpha v - z\| \\
&\leq \|x - \alpha v - z\|^2 + \alpha^2 \delta_f^2 + 2\alpha \delta_f (K_0 + \bar{K} + L_0 + 1 + \Delta r^{-1}) \\
&\leq \|x - z\|^2 - 2\alpha \langle x - z, v \rangle + \alpha^2 \|v\|^2 \\
&\quad + \alpha^2 \delta_f^2 + 2\alpha \delta_f (K_0 + \bar{K} + L_0 + 1 + \Delta r^{-1}) \\
&\leq \|x - z\|^2 - 2\alpha \langle x - z, v \rangle + \alpha^2 (L_0 + \Delta r^{-1})^2 \\
&\quad + \alpha^2 \delta_f^2 + 2\alpha \delta_f (K_0 + \bar{K} + L_0 + 1 + \Delta r^{-1}). \tag{3.117}
\end{aligned}
$$

In view of (3.112),

$$\langle v, z - x \rangle \leq f(z) - f(x) + \Delta. \tag{3.118}$$

By (3.105), (3.111), and (3.116),

$$\|x - \alpha\xi - z\| \leq K_0 + \bar{K} + 2 + L_0 + \Delta r^{-1}. \tag{3.119}$$

It follows from (3.117) that

$$\|x - \alpha\xi - z\|^2 \leq \|x - z\|^2$$
$$+ 2\alpha(f(z) - f(x)) + 2\alpha\Delta + \alpha^2(L_0 + \Delta r^{-1})^2$$
$$+ 2\alpha\delta_f(K_0 + \bar{K} + L_0 + 2 + \Delta r^{-1}). \tag{3.120}$$

It follows from (3.56), (3.57), (3.109), (3.119), and (3.120) that

$$\|y - z\|^2 = \|y - P(x - \alpha\xi) + P(x - \alpha\xi) - z\|^2$$
$$\leq \|y - P(x - \alpha\xi)\|^2 + 2\|y - P(x - \alpha\xi)\|\|P(x - \alpha\xi) - z\|$$
$$+ \|P(x - \alpha\xi) - z\|^2$$
$$\leq \delta_C^2 + 2\delta_C(K_0 + 2 + \bar{K} + L_0 + \Delta r^{-1}) + \|P(x - \alpha\xi) - z\|^2$$
$$\leq 2\delta_C(K_0 + 3 + \bar{K} + L_0 + \Delta r^{-1}) + \|x - z\|^2$$
$$+ 2\alpha(f(z) - f(x)) + 2\alpha\Delta + \alpha^2(L_0 + \Delta r^{-1})^2$$
$$+ 2\alpha\delta_f(K_0 + \bar{K} + L_0 + 2 + \Delta r^{-1}).$$

This implies that

$$2\alpha(f(x) - f(z)) \leq \|x - z\|^2 - \|y - z\|^2$$
$$+ 2\delta_C(K_0 + 3 + \bar{K} + L_0 + \Delta r^{-1}) + 2\alpha\Delta + \alpha^2(L_0 + \Delta r^{-1})^2$$
$$+ 2\alpha\delta_f(K_0 + \bar{K} + L_0 + 2 + \Delta r^{-1}).$$

Lemma 3.11 is proved.

3.8   Proof of *Theorem 3.6

Fix

$$\bar{x} \in C_{min}. \tag{3.121}$$

By (3.52), (3.54), and (3.121),

$$\|\bar{x}\| \leq \bar{K}. \tag{3.122}$$

In view of (3.64) and (3.122),

$$\|x_0 - \bar{x}\| \leq \bar{K} + K_1. \tag{3.123}$$

By induction we show that for all $i = 0, \ldots, T$,

$$\|x_i - \bar{x}\| \leq \bar{K} + K_1. \tag{3.124}$$

In view of (3.123), inequality (3.124) holds for $i = 0$. Assume that $i \in \{0, \ldots, T - 1\}$ and that (3.124) is valid. There are two cases:

$$f(x_i) \leq \inf(f, C) + 4; \tag{3.125}$$

$$f(x_i) > \inf(f, C) + 4. \tag{3.126}$$

Assume that (3.125) holds. In view of (3.52)–(3.54) and (3.125),

$$\|x_i\| \leq \bar{K}. \tag{3.127}$$

Lemma 3.8, (3.55), and (3.127) imply that

$$\partial f(x_i) \subset B_X(0, \bar{L}). \tag{3.128}$$

By (3.66) and (3.128),

$$\|\xi_i\| \leq \bar{L} + 1. \tag{3.129}$$

It follows from (3.56), (3.57), (3.63), (3.67), (3.121), (3.122), (3.127), and (3.129) that

$$\|x_{i+1} - \bar{x}\| \leq \|x_{i+1} - P_i(x_i - \alpha\xi_i)\| + \|P_i(x_i - \alpha\xi_i) - \bar{x}\|$$
$$\leq \delta_C + \|x_i - \alpha\xi_i - \bar{x}\|$$
$$\leq \delta_C + 2\bar{K} + \alpha(\bar{L} + 1) \leq 2\bar{K} + 2 \leq K_1 + \bar{K}.$$

Assume that (3.126) holds. In view of (3.66), (3.121), (3.122), and (3.124), we apply Lemma 3.10 with

$$P = P_i, \ K_0 = 3K_1, \ L_0 = L_1, \ \xi = \xi_i, \ x = x_i, \ y = x_{i+1}, \ z = \bar{x}$$

and obtain that

$$2\alpha(f(x_i) - f(\bar{x})) \leq \|x_i - \bar{x}\|^2 - \|x_{i+1} - \bar{x}\|^2$$
$$+ 2\delta_C(\bar{K} + 3K_1 + L_1 + 3) + L_1^2\alpha^2 + 2\alpha\delta_f(3K_1 + \bar{K} + L_1 + 2).$$

Together with (3.60), (3.61), and (3.126), this implies that

$$\|x_{i+1} - \bar{x}\|^2 \leq \|x_i - \bar{x}\|^2 - 8\alpha + 2\delta_C(\bar{K} + 3K_1 + L_1 + 3)$$

$$+ L_1^2 \alpha^2 + 2\alpha \delta_f (3K_1 + \bar{K} + L_1 + 2)$$

$$\leq \|x_i - \bar{x}\|^2 - 7\alpha + 4\alpha,$$

$$\|x_{i+1} - \bar{x}\| \leq \|x_i - \bar{x}\| \leq \bar{K} + K_1.$$

Thus in both cases

$$\|x_{i+1} - \bar{x}\| \leq \bar{K} + K_1$$

and the assumption made for i also holds for $i + 1$ too. Therefore (3.124) holds for all $i = 0, \ldots, T$.

In view of (3.122) and (3.124), for all $t = 0, \ldots, T$,

$$\|x_i\| \leq 2\bar{K} + K_1. \tag{3.130}$$

Let $i \in \{0, \ldots, T - 1\}$. In view of (3.58), (3.59), (3.66), and (3.130), we apply Lemma 3.10 with $P = P_i$, $K_0 = 3K_1$, $L_0 = L_1$, $x = x_i$, $y = x_{i+1}$, $z = \bar{x}$, $\xi = \xi_i$ and obtain that

$$2\alpha (f(x_i) - f(\bar{x})) \leq \|x_i - \bar{x}\|^2 - \|x_{i+1} - \bar{x}\|^2$$

$$+ 2\delta_C (\bar{K} + 3K_1 + L_1 + 3) + \alpha^2 L_1^2 + 2\alpha \delta_f (3K_1 + \bar{K} + L_1 + 2). \tag{3.131}$$

By (3.131),

$$\sum_{i=0}^{T-1} \alpha (f(x_i) - f(\bar{x})) \leq 2^{-1} \sum_{i=0}^{T-1} (\|x_i - \bar{x}\|^2 - \|x_{i+1} - \bar{x}\|^2)$$

$$+ T\delta_C (\bar{K} + 3K_1 + L_1 + 3) + T\alpha^2 L_1^2$$

$$+ T\alpha \delta_f (3K_1 + \bar{K} + L_1 + 2).$$

Together with (3.121) and (3.123), the relation above implies that

$$\min\{f(x_t) : \ t = 0, \ldots, T - 1\} - \inf(f, C), \ \ f(T^{-1} \sum_{i=0}^{T-1} x_i) - \inf(f, C)$$

$$\leq T^{-1} \sum_{i=0}^{T-1} f(x_i) - \inf(f, C)$$

$$\leq (2T\alpha)^{-1}(K_1 + \bar{K})^2 + L_1^2\alpha + \alpha^{-1}\delta_C(\bar{K} + 3K_1 + L_1 + 3)$$
$$+ \delta_f(3K_1 + \bar{K} + L_1 + 2).$$

Theorem 3.6 is proved.

3.9   Proof of Theorem 3.7

Fix

$$\bar{x} \in C_{min}. \tag{3.132}$$

By (3.52)–(3.54) and (3.132),

$$\|\bar{x}\| \leq \bar{K}. \tag{3.133}$$

In view of (3.75) and (3.133),

$$\|x_0 - \bar{x}\| \leq \bar{K} + K_1. \tag{3.134}$$

By induction we show that for all $i = 0, \ldots, T$,

$$\|x_i - \bar{x}\| \leq \bar{K} + K_1. \tag{3.135}$$

In view of (3.134), inequality (3.135) holds for $i = 0$. Assume that $i \in \{0, \ldots, T - 1\}$ and that (3.135) is valid. There are two cases:

$$f(x_i) \leq \inf(f, C) + 4; \tag{3.136}$$

$$f(x_i) > \inf(f, C) + 4. \tag{3.137}$$

Assume that (3.136) holds. In view of (3.52)–(3.54) and (3.136),

$$\|x_i\| \leq \bar{K}. \tag{3.138}$$

It follows from (3.68), (3.76), and (3.78) that

$$B_X(x_i, r_0/2) \subset U. \tag{3.139}$$

Lemma 3.9, (3.55), (3.138), and (3.139) imply that

$$\partial_\Delta f(x_i) \subset B_X(0, \bar{L} + 2\Delta r_0^{-1}). \tag{3.140}$$

By (3.77) and (3.140),

$$\|\xi_i\| \leq \bar{L} + 1 + 2\Delta r_0^{-1}. \tag{3.141}$$

It follows from (3.56), (3.57), (3.70), (3.73), (3.78), (3.133), (3.138), and (3.141) that

$$\begin{aligned}
\|x_{i+1} - \bar{x}\| &\leq \|x_{i+1} - P_i(x_i - \alpha\xi_i)\| + \|P_i(x_i - \alpha\xi_i) - \bar{x}\| \\
&\leq \delta_C + \|x_i - \alpha\xi_i - \bar{x}\| \\
&\leq \delta_C + 2\bar{K} + \alpha(\bar{L} + 1 + 2\Delta r_0^{-1}) \leq 2\bar{K} + 1 + \alpha(\bar{L} + 3) \\
&\leq 2K + 2 \leq \bar{K} + K_1.
\end{aligned}$$

Assume that (3.137) holds. In view of (3.133) and (3.135),

$$\|x_i\| \leq 2\bar{K} + K_1. \tag{3.142}$$

By (3.68), (3.70), (3.76), and (3.78), equation (3.139) is true.

In view of (3.69), (3.70), (3.73), (3.74), (3.77), (3.78), and (3.142), we apply Lemma 3.11 with

$$P = P_i, \ K_0 = 2\bar{K} + K_1, \ L_0 = L_1, \ r = r_0/2, \ \xi = \xi_i, \ x = x_i, \ y = x_{i+1}, \ z = \bar{x}$$

and obtain that

$$\begin{aligned}
2\alpha(f(x_i) - f(\bar{x})) &\leq \|x_i - \bar{x}\|^2 - \|x_{i+1} - \bar{x}\|^2 \\
&+ 2\delta_C(3\bar{K} + K_1 + L_1 + 3 + 2\Delta r_0^{-1}) + 2\alpha\Delta \\
&+ (L_1 + 2\Delta r_0^{-1})^2\alpha^2 + 2\alpha\delta_f(K_1 + 3\bar{K} + L_1 + 2 + 2\Delta r_0^{-1}). \tag{3.143}
\end{aligned}$$

Together with (3.70)–(3.72), (3.135), and (3.127), this implies that

$$\begin{aligned}
\|x_{i+1} - \bar{x}\|^2 &\leq \|x_i - \bar{x}\|^2 - 8\alpha + 2\delta_C(3\bar{K} + K_1 + L_1 + 3 + 2\Delta r_0^{-1}) \\
&+ 2\alpha\Delta + (L_1 + 2\Delta r_0^{-1})^2\alpha^2 + 2\alpha\delta_f(K_1 + 3\bar{K} + L_1 + 2 + 2\Delta r_0^{-1}) \\
&\leq \|x_i - \bar{x}\|^2 - 8\alpha + 7\alpha < \|x_i - \bar{x}\|^2, \\
\|x_{i+1} - \bar{x}\| &\leq \|x_i - \bar{x}\| \leq \bar{K} + K_1.
\end{aligned}$$

Thus in both cases

$$\|x_{i+1} - \bar{x}\| \leq \bar{K} + K_1$$

and the assumption made for i also holds for $i + 1$ too. Therefore (3.135) holds for all $i = 0, \ldots, T$. In view of (3.133) and (3.135), for all $t = 0, \ldots, T$,

$$\|x_i\| \le 2\bar{K} + K_1. \tag{3.144}$$

Let $i \in \{0, \ldots, T-1\}$. By (3.68), (3.70), (3.76), and (3.78), equation (3.139) is true. In view of (3.69), (3.74), (3.77), (3.78), (3.139), and (3.143), we apply Lemma 3.11 with $P = P_i$, $K_0 = K_1 + 2\bar{K}$, $L_0 = L_1$, $r = r_0/2$, $x = x_i$, $y = x_{i+1}$, $z = \bar{x}$, $\xi = \xi_i$ and obtain that (3.143) is true. By (3.70) and (3.143),

$$\sum_{i=0}^{T-1} \alpha(f(x_i) - f(\bar{x})) \le 2^{-1} \sum_{i=0}^{T-1} (\|x_i - \bar{x}\|^2 - \|x_{i+1} - \bar{x}\|^2)$$
$$+ T\delta_C(3\bar{K} + K_1 + L_1 + 2) + \alpha\Delta T + T\alpha^2(L_1 + 2)^2$$
$$+ T\alpha\delta_f(K_1 + 3\bar{K} + L_1 + 4).$$

Together with (3.135) the relation above implies that

$$\min\{f(x_t) : t = 0, \ldots, T - 1\} - \inf(f, C), \ f\left(T^{-1} \sum_{i=0}^{T-1} x_i\right) - \inf(f, C)$$

$$\le T^{-1} \sum_{i=0}^{T-1} f(x_i) - \inf(f, C)$$

$$\le (2T\alpha)^{-1}(K_1 + \bar{K})^2 + \alpha^{-1}\delta_C(3\bar{K} + K_1 + L_1 + 2)$$
$$+ \Delta + \alpha(L_1 + 2)^2 + \delta_f(K_1 + 3\bar{K} + L_1 + 2).$$

Theorem 3.7 is proved.

Chapter 4
Zero-Sum Games with Two Players

In this chapter we study an extension of the projected subgradient method for zero-sum games with two players under the presence of computational errors. In our recent research [77], we show that our algorithm generates a good approximate solution, if all the computational errors are bounded from above by a small positive constant. Moreover, if we know computational errors for our algorithm, we find out what an approximate solution can be obtained and how many iterates one needs for this. In this chapter we generalize these results for an extension of the projected subgradient method, when instead of the projection on a feasible set it is used a quasi-nonexpansive retraction on this set.

4.1 Preliminaries and an Auxiliary Result

Let $(X, \langle \cdot, \cdot \rangle)$, $(Y, \langle \cdot, \cdot \rangle)$ be Hilbert spaces equipped with the complete norms $\| \cdot \|$ which are induced by their inner products. Let C be a nonempty closed convex subset of X, D be a nonempty closed convex subset of Y, U be an open convex subset of X, and V be an open convex subset of Y such that

$$C \subset U, \ D \subset V \tag{4.1}$$

and let a function $f : U \times V \to R^1$ possess the following properties:

(i) for each $v \in V$, the function $f(\cdot, v) : U \to R^1$ is convex;
(ii) for each $u \in U$, the function $f(u, \cdot) : V \to R^1$ is concave.

Assume that a function $\phi : R^1 \to [0, \infty)$ is bounded on all bounded sets and that positive numbers M_1, M_2, L_1, L_2 satisfy

$$C \subset B_X(0, M_1),$$

A. J. Zaslavski, *The Projected Subgradient Algorithm in Convex Optimization*,
SpringerBriefs in Optimization, https://doi.org/10.1007/978-3-030-60300-7_4

$$D \subset B_Y(0, M_2), \tag{4.2}$$

$$|f(u_1, v) - f(u_2, v)| \le L_1 \|u_1 - u_2\|$$

$$\text{for all } v \in V \text{ and all } u_1, u_2 \in U, \tag{4.3}$$

$$|f(u, v_1) - f(u, v_2)| \le L_2 \|v_1 - v_2\|$$

$$\text{for all } u \in U \text{ and all } v_1, v_2 \in V. \tag{4.4}$$

Let

$$x_* \in C \text{ and } y_* \in D \tag{4.5}$$

satisfy

$$f(x_*, y) \le f(x_*, y_*) \le f(x, y_*) \tag{4.6}$$

for each $x \in C$ and each $y \in D$.

The following result was obtained in [77].

Proposition 4.1 *Let T be a natural number, δ_C, $\delta_D \in (0, 1]$, $\{a_t\}_{t=0}^T \subset (0, \infty)$ and let $\{b_{t,1}\}_{t=0}^T$, $\{b_{t,2}\}_{t=0}^T \subset (0, \infty)$. Assume that $\{x_t\}_{t=0}^{T+1} \subset U$, $\{y_t\}_{t=0}^{T+1} \subset V$, for each $t \in \{0, \dots, T+1\}$,*

$$B_X(x_t, \delta_C) \cap C \ne \emptyset,$$

$$B_Y(y_t, \delta_D) \cap D \ne \emptyset,$$

for each $z \in C$ and each $t \in \{0, \dots, T\}$,

$$a_t(f(x_t, y_t) - f(z, y_t))$$

$$\le \phi(\|z - x_t\|) - \phi(\|z - x_{t+1}\|) + b_{t,1}$$

and that for each $v \in D$ and each $t \in \{0, \dots, T\}$,

$$a_t(f(x_t, v) - f(x_t, y_t))$$

$$\le \phi(\|v - y_t\|) - \phi(\|v - y_{t+1}\|) + b_{t,2}.$$

Let

$$\widehat{x}_T = \left(\sum_{i=0}^{T} a_i \right)^{-1} \sum_{t=0}^{T} a_t x_t,$$

$$\widehat{y}_T = \left(\sum_{i=0}^{T} a_i \right)^{-1} \sum_{t=0}^{T} a_t y_t.$$

Then

$$B_X(\widehat{x}_T, \delta_C) \cap C \neq \emptyset, \ \ B_Y(\widehat{y}_T, \delta_D) \cap D \neq \emptyset,$$

$$\left| \left(\sum_{t=0}^{T} a_t \right)^{-1} \sum_{t=0}^{T} a_t f(x_t, y_t) - f(x_*, y_*) \right|$$

$$\leq \left(\sum_{t=0}^{T} a_t \right)^{-1} \max \left\{ \sum_{t=0}^{T} b_{t,1}, \ \sum_{t=0}^{T} b_{t,2} \right\}$$

$$+ \max\{ L_1 \delta_C, \ L_2 \delta_D \}$$

$$+ \left(\sum_{t=0}^{T} a_t \right)^{-1} \sup\{\phi(s) : \ s \in [0, \max\{2M_1, \ 2M_2\} + 1]\},$$

$$\left| f(\widehat{x}_T, \widehat{y}_T) - \left(\sum_{t=0}^{T} a_t \right)^{-1} \sum_{t=0}^{T} a_t f(x_t, y_t) \right|$$

$$\leq \left(\sum_{t=0}^{T} a_t \right)^{-1} \sup\{\phi(s) : \ s \in [0, \max\{2M_1, \ 2M_2\} + 1]\}$$

$$+ \left(\sum_{t=0}^{T} a_t \right)^{-1} \max \left\{ \sum_{t=0}^{T} b_{t,1}, \ \sum_{t=0}^{T} b_{t,2} \right\}$$

$$+ \max\{ L_1 \delta_C, \ L_2 \delta_D \}$$

and for each $z \in C$ and each $v \in D$,

$$f(z, \widehat{y}_T) \geq f(\widehat{x}_T, \widehat{y}_T)$$

$$-2 \left(\sum_{t=0}^{T} a_t \right)^{-1} \sup\{\phi(s) : \ s \in [0, \max\{2M_1, \ 2M_2\} + 1]\}$$

$$-2 \left(\sum_{t=0}^{T} a_t \right)^{-1} \max \left\{ \sum_{t=0}^{T} b_{t,1}, \ \sum_{t=0}^{T} b_{t,2} \right\}$$

$$- \max\{L_1 \delta_C, \ L_2 \delta_D\},$$

$$f(\widehat{x}_T, v) \le f(\widehat{x}_T, \widehat{y}_T)$$

$$+2 \left(\sum_{t=0}^{T} a_t \right)^{-1} \sup\{\phi(s) : \ s \in [0, \max\{2M_1, \ 2M_2\} + 1]\}$$

$$+2 \left(\sum_{t=0}^{T} a_t \right)^{-1} \max \left\{ \sum_{t=0}^{T} b_{t,1}, \ \sum_{t=0}^{T} b_{t,2} \right\}$$

$$+ \max\{L_1 \delta_C, \ L_2 \delta_D\}.$$

The following corollary was obtained in [77].

Corollary 4.2 *Suppose that all the assumptions of Proposition 4.1 hold and that*

$$\tilde{x} \in C, \ \tilde{y} \in D$$

satisfy

$$\|\widehat{x}_T - \tilde{x}\| \le \delta_C, \ \|\widehat{y}_T - \tilde{y}\| \le \delta_D.$$

Then

$$|f(\tilde{x}, \tilde{y}) - f(\widehat{x}_T, \widehat{y}_T)| \le L_1 \delta_C + L_2 \delta_D$$

and for each $z \in C$ and each $v \in D$,

$$f(z, \tilde{y}) \ge f(\tilde{x}, \tilde{y})$$

$$-2 \left(\sum_{t=0}^{T} a_t \right)^{-1} \sup\{\phi(s) : \ s \in [0, \max\{2M_1, \ 2M_2\} + 1]\}$$

$$-2 \left(\sum_{t=0}^{T} a_t \right)^{-1} \max \left\{ \sum_{t=0}^{T} b_{t,1}, \ \sum_{t=0}^{T} b_{t,2} \right\} - 4 \max\{L_1 \delta_C, \ L_2 \delta_D\}$$

and

$$f(\tilde{x}, v) \le f(\tilde{x}, \tilde{y})$$

$$+2 \left(\sum_{t=0}^{T} a_t \right)^{-1} \sup\{\phi(s) : s \in [0, \max\{2M_1, 2M_2\} + 1]\}$$

$$+2 \left(\sum_{t=0}^{T} a_t \right)^{-1} \max \left\{ \sum_{t=0}^{T} b_{t,1}, \sum_{t=0}^{T} b_{t,2} \right\} + 4 \max\{L_1 \delta_C, L_2 \delta_D\}.$$

4.2   Zero-Sum Games on Bounded Sets

Let $(X, \langle \cdot, \cdot \rangle)$, $(Y, \langle \cdot, \cdot \rangle)$ be Hilbert spaces equipped with the complete norms $\| \cdot \|$ which are induced by their inner products. Let C be a nonempty closed convex subset of X, D be a nonempty closed convex subset of Y, U be an open convex subset of X, and V be an open convex subset of Y such that

$$C \subset U, \ D \subset V.$$

For each concave function $g : V \to R^1$, each $x \in V$, and each $\epsilon > 0$, set

$$\partial g(x) = \{l \in Y : \langle l, y - x \rangle \geq g(y) - g(x) \text{ for all } y \in V\}, \tag{4.7}$$

$$\partial_\epsilon g(x) = \{l \in Y : \langle l, y - x \rangle + \epsilon \geq g(y) - g(x) \text{ for all } y \in V\}. \tag{4.8}$$

Clearly, for each $x \in V$ and each $\epsilon > 0$,

$$\partial g(x) = -(\partial(-g)(x)), \tag{4.9}$$

$$\partial_\epsilon g(x) = -(\partial_\epsilon(-g)(x)). \tag{4.10}$$

Suppose that there exist $L_1, L_2, M_1, M_2 > 0$ such that

$$C \subset B_X(0, M_1), \ D \subset B_Y(0, M_2), \tag{4.11}$$

a function $f : U \times V \to R^1$ possesses the following properties:

(i) for each $v \in V$, the function $f(\cdot, v) : U \to R^1$ is convex;
(ii) for each $u \in U$, the function $f(u, \cdot) : V \to R^1$ is concave,

for each $v \in V$,

$$|f(u_1, v) - f(u_2, v)| \leq L_1 \|u_1 - u_2\|$$

$$\text{for all } u_1, u_2 \in U \tag{4.12}$$

and that for each $u \in U$,

$$|f(u, v_1) - f(u, v_2)| \leq L_2 \|v_1 - v_2\|$$

$$\text{for all } v_1, v_2 \in V. \tag{4.13}$$

For each $(\xi, \eta) \in U \times V$ and each $\epsilon > 0$, set

$$\partial_x f(\xi, \eta) = \{l \in X :$$
$$f(y, \eta) - f(\xi, \eta) \geq \langle l, y - \xi \rangle \text{ for all } y \in U\}, \tag{4.14}$$

$$\partial_y f(\xi, \eta) = \{l \in Y :$$
$$\langle l, y - \eta \rangle \geq f(\xi, y) - f(\xi, \eta) \text{ for all } y \in V\}, \tag{4.15}$$

$$\partial_{x,\epsilon} f(\xi, \eta) = \{l \in X :$$
$$f(y, \eta) - f(\xi, \eta) + \epsilon \geq \langle l, y - \xi \rangle \text{ for all } y \in U\}, \tag{4.16}$$

$$\partial_{y,\epsilon} f(\xi, \eta) = \{l \in Y :$$
$$\langle l, y - \eta \rangle + \epsilon \geq f(\xi, y) - f(\xi, \eta) \text{ for all } y \in V\}. \tag{4.17}$$

In view of properties (i) and (ii), (4.12), and (4.13), for each $\xi \in U$ and each $\eta \in V$,

$$\emptyset \neq \partial_x f(\xi, \eta) \subset B_X(0, L_1), \tag{4.18}$$

$$\emptyset \neq \partial_y f(\xi, \eta) \subset B_Y(0, L_2). \tag{4.19}$$

Let

$$x_* \in C \text{ and } y_* \in D \tag{4.20}$$

satisfy

$$f(x_*, y) \leq f(x_*, y_*) \leq f(x, y_*) \tag{4.21}$$

for each $x \in C$ and each $y \in D$.

Denote by $\mathcal{M}_U$ the set of all mappings $P : X \to X$ such that

$$Px = x, \quad x \in C,$$

$$\|Px - z\| \leq \|x - z\| \text{ for all } x \in X \text{ and all } z \in C$$

and by $\mathcal{M}_V$ the set of all mappings $P : Y \to Y$ such that

$$Py = y, \ y \in D,$$

$$\|Py - z\| \leq \|y - z\| \text{ for all } y \in Y \text{ and all } z \in C.$$

Let $\delta_{f,1}, \delta_{f,2}, \delta_C, \delta_D \in (0, 1]$ and $\{\alpha_k\}_{k=0}^{\infty} \subset (0, \infty)$.
Let us describe our algorithm.

Subgradient Projection Algorithm for Zero-Sum Games
Initialization: select arbitrary $x_0 \in U$ and $y_0 \in V$.
Iterative step: given the current iteration vectors $x_t \in U$ and $y_t \in V$, calculate

$$\xi_t \in \partial_x f(x_t, y_t) + B_X(0, \delta_{f,1}),$$

$$\eta_t \in \partial_y f(x_t, y_t) + B_Y(0, \delta_{f,2})$$

and the next pair of iteration vectors $x_{t+1} \in U$, $y_{t+1} \in V$ such that

$$\|x_{t+1} - P_t(x_t - \alpha_t \xi_t)\| \leq \delta_C,$$

$$\|y_{t+1} - Q_t(y_t + \alpha_t \eta_t)\| \leq \delta_D,$$

where $P_t \in \mathcal{M}_U$, $Q_t \in \mathcal{M}_V$.
In this chapter we prove the following result.

Theorem 4.3 *Let* $\delta_{f,1}, \delta_{f,2}, \delta_C, \delta_D \in (0, 1]$, $\{\alpha_k\}_{k=0}^{\infty} \subset (0, \infty)$,

$$\{P_t\}_{t=0}^{\infty} \subset \mathcal{M}_U, \ P_t(X) = C, \ t = 0, 1, \ldots, \tag{4.22}$$

$$\{Q_t\}_{t=0}^{\infty} \subset \mathcal{M}_V, \ Q_t(Y) = D, \ t = 0, 1, \ldots. \tag{4.23}$$

Assume that $\{x_t\}_{t=0}^{\infty} \subset U$, $\{y_t\}_{t=0}^{\infty} \subset V$, $\{\xi_t\}_{t=0}^{\infty} \subset X$, $\{\eta_t\}_{t=0}^{\infty} \subset Y$,

$$B_X(x_0, \delta_C) \cap C \neq \emptyset, \ B_Y(y_0, \delta_D) \cap D \neq \emptyset \tag{4.24}$$

and that for each integer $t \geq 0$,

$$\xi_t \in \partial_x f(x_t, y_t) + B_X(0, \delta_{f,1}), \tag{4.25}$$

$$\eta_t \in \partial_y f(x_t, y_t) + B_Y(0, \delta_{f,2}), \tag{4.26}$$

$$\|x_{t+1} - P_t(x_t - \alpha_t \xi_t)\| \leq \delta_C \tag{4.27}$$

and

$$\|y_{t+1} - Q_t(y_t + \alpha_t \eta_t)\| \leq \delta_D. \tag{4.28}$$

For each integer $t \geq 0$ set

$$b_{t,1} = \alpha_t^2 L_1^2 + \delta_C(2M_1 + L_1 + 3) + \alpha_t \delta_{f,1}(2M_1 + L_1 + 2), \tag{4.29}$$

$$b_{t,2} = \alpha_t^2 L_2^2 + \delta_D(2M_2 + L_2 + 3) + \alpha_t \delta_{f,2}(2M_2 + L_2 + 2). \tag{4.30}$$

Let for each natural number T,

$$\widehat{x}_T = \left(\sum_{i=0}^{T} \alpha_t\right)^{-1} \sum_{t=0}^{T} \alpha_t x_t, \tag{4.31}$$

$$\widehat{y}_T = \left(\sum_{i=0}^{T} \alpha_t\right)^{-1} \sum_{t=0}^{T} \alpha_t y_t. \tag{4.32}$$

Then for each natural number T,

$$B_X(\widehat{x}_T, \delta_C) \cap C \neq \emptyset, \quad B_Y(\widehat{y}_T, \delta_D) \cap D \neq \emptyset,$$

$$\left| \left(\sum_{t=0}^{T} \alpha_t\right)^{-1} \sum_{t=0}^{T} \alpha_t f(x_t, y_t) - f(x_*, y_*) \right|$$

$$\leq \left(\sum_{t=0}^{T} \alpha_t\right)^{-1} \max\left\{ \sum_{t=0}^{T} b_{t,1}, \sum_{t=0}^{T} b_{t,2} \right\} + \max\{L_1 \delta_C, \ L_2 \delta_D\}$$

$$+ \left(2\sum_{t=0}^{T} \alpha_t\right)^{-1} \max\{(2M_1, \ 2M_2\} + 1)^2,$$

$$\left| f(\widehat{x}_T, \widehat{y}_T) - \left(\sum_{t=0}^{T} \alpha_t\right)^{-1} \sum_{t=0}^{T} \alpha_t f(x_t, y_t) \right|$$

$$\leq \left(2\sum_{t=0}^{T} \alpha_t\right)^{-1} (\max\{2M_1, 2M_2\} + 1)^2 + \max\{L_1 \delta_C, \ L_2 \delta_D\}$$

$$+ \left(\sum_{t=0}^{T} \alpha_t\right)^{-1} \max\left\{ \sum_{t=0}^{T} b_{t,1}, \sum_{t=0}^{T} b_{t,2} \right\}$$

and for each $z \in C$ and each $v \in D$,

$$f(z, \widehat{y}_T) \geq f(\widehat{x}_T, \widehat{y}_T)$$

$$- \left(\sum_{t=0}^{T} \alpha_t \right)^{-1} (\max\{2M_1, 2M_2\} + 1)^2 - \max\{L_1\delta_C, \ L_2\delta_D\}$$

$$-2 \left(\sum_{t=0}^{T} \alpha_t \right)^{-1} \max\left\{ \sum_{t=0}^{T} b_{t,1}, \sum_{t=0}^{T} b_{t,2} \right\}$$

$$f(\widehat{x}_T, v) \leq f(\widehat{x}_T, \widehat{y}_T)$$

$$+ \left(\sum_{t=0}^{T} \alpha_t \right)^{-1} (\max\{2M_1, 2M_2\} + 1)^2 + \max\{L_1\delta_C, \ L_2\delta_D\}$$

$$+2 \left(\sum_{t=0}^{T} \alpha_t \right)^{-1} \max\left\{ \sum_{t=0}^{T} b_{t,1}, \sum_{t=0}^{T} b_{t,2} \right\}.$$

Proof By (4.11), (4.22)–(4.24), (4.27), and (4.28), for all integers $t \geq 0$,

$$\|x_t\| \leq M_1 + 1, \ \|y_t\| \leq M_2 + 1. \tag{4.33}$$

Let $t \geq 0$ be an integer. Applying Lemma 3.4 with

$$P = P_t, \ \delta_f = \delta_{f,1}, \ \alpha = \alpha_t, \ x = x_t, \ f = f(\cdot, y_t), \ \xi = \xi_t, \ y = x_{t+1},$$

we obtain that for each $z \in C$,

$$\alpha_t (f(x_t, y_t) - f(z, y_t)) \leq 2^{-1} \|z - x_t\|^2 - 2^{-1} \|z - x_{t+1}\|^2$$
$$+\alpha_t^2 L_1^2 + \delta_C (2M_1 + L_1 + 3) + \alpha_t \delta_{f,1}(2M_1 + L_1 + 2)$$
$$\leq 2^{-1} \|z - x_t\|^2 - 2^{-1} \|z - x_{t+1}\|^2 + b_{t,1}.$$

Applying Lemma 3.4 with

$$P = Q_t, \ \alpha = \alpha_t, \ x = y_t, \ f = -f(x_t, \cdot), \ \xi = -\eta_t, \ y = y_{t+1}, \ \delta_f = \delta_{f,2}$$

we obtain that for each $v \in D$,

$$\alpha_t (f(x_t, v) - f(x_t, y_t)) \leq 2^{-1} \|v - y_t\|^2 - 2^{-1} \|v - y_{t+1}\|^2$$
$$+\delta_D (2M_2 + L_2 + 3) + \alpha_t \delta_{f,2}(2M_2 + L_2 + 2) + \alpha_t^2 L_2^2$$

$$\leq 2^{-1} \|v - y_t\|^2 - 2^{-1} \|v - y_{t+1}\|^2 + b_{t,2}.$$

Define

$$\phi(s) = 2^{-1} s^2, \ s \in R^1.$$

It is easy to see that all the assumptions of Proposition 4.1 hold and it implies Theorem 4.3.

Theorem 4.4 *Let* $r_1, r_2 > 0$,

$$B_X(z, 2r_1) \subset U \text{ for all } z \in C, \tag{4.34}$$

$$B_Y(u, 2r_2) \subset V \text{ for all } u \in D, \tag{4.35}$$

$\Delta_1, \Delta_2 > 0$, $\delta_{f,1}, \delta_{f,2}, \delta_C, \delta_D \in (0, 1]$,

$$\delta_C \leq r_1, \ \delta_D \leq r_2, \tag{4.36}$$

and $\{\alpha_t\}_{k=0}^{\infty} \subset (0, 1]$,

$$\{P_t\}_{t=0}^{\infty} \subset \mathcal{M}_U, \ \{Q_t\}_{t=0}^{\infty} \subset \mathcal{M}_V, \tag{4.37}$$

$$P_t(X) = C, \ t = 0, 1, \dots, \ Q_t(Y) = D, \ t = 0, 1, \dots. \tag{4.38}$$

Assume that $\{x_t\}_{t=0}^{\infty} \subset U$, $\{y_t\}_{t=0}^{\infty} \subset V$, $\{\xi_t\}_{t=0}^{\infty} \subset X$, $\{\eta_t\}_{t=0}^{\infty} \subset Y$,

$$B_X(x_0, \delta_C) \cap C \neq \emptyset, \ B_Y(y_0, \delta_D) \cap D \neq \emptyset \tag{4.39}$$

and that for each integer $t \geq 0$,

$$B_X(\xi_t, \delta_{f,1}) \cap \partial_{x, \Delta_1} f(x_t, y_t) \neq \emptyset,$$

$$B_Y(\eta_t, \delta_{f,2}) \cap \partial_{y, \Delta_2} f(x_t, y_t) \neq \emptyset, \tag{4.40}$$

$$\|x_{t+1} - P_t(x_t - a_t \xi_t)\| \leq \delta_C$$

and

$$\|y_{t+1} - Q_t(y_t + a_t \eta_t)\| \leq \delta_D. \tag{4.41}$$

For each integer $t \geq 0$ *set*

$$b_{t,1} = \alpha_t \Delta_1 + 2^{-1} \alpha_t^2 (L_1 + \Delta_1 r_1^{-1})$$

$$+\delta_C(2M_1 + L_1 + 3 + \Delta_1 r_1^{-1}) + \alpha_t \delta_{f,1}(2M_1 + L_1 + 2 + \Delta_1 r_1^{-1}),$$

$$b_{t,2} = \alpha_t \Delta_2 + 2^{-1}\alpha_t^2(L_2 + \Delta_2 r_2^{-1})$$

$$+\delta_D(2M_2 + L_2 + 3 + \Delta_2 r_2^{-1}) + \alpha_t \delta_{f,2}(2M_2 + L_2 + 2 + \Delta_2 r_2^{-1}).$$

Let for each natural number T

$$\widehat{x}_T = \left(\sum_{i=0}^{T} \alpha_t\right)^{-1} \sum_{t=0}^{T} \alpha_t x_t,$$

$$\widehat{y}_T = \left(\sum_{i=0}^{T} \alpha_t\right)^{-1} \sum_{t=0}^{T} \alpha_t y_t.$$

Then for each natural number T,

$$B_X(\widehat{x}_T, \delta_C) \cap C \neq \emptyset, \quad B_Y(\widehat{y}_T, \delta_D) \cap D \neq \emptyset,$$

$$\left| \left(\sum_{t=0}^{T} \alpha_t\right)^{-1} \sum_{t=0}^{T} \alpha_t f(x_t, y_t) - f(x_*, y_*) \right|$$

$$\leq \left(\sum_{t=0}^{T} \alpha_t\right)^{-1} \max\left\{\sum_{t=0}^{T} b_{t,1}, \sum_{t=0}^{T} b_{t,2}\right\} + \max\{L_1\delta_C, \ L_2\delta_D\}$$

$$+ \left(2\sum_{t=0}^{T} \alpha_t\right)^{-1} \max\{(2M_1, \ 2M_2) + 1)^2,$$

$$\left| f(\widehat{x}_T, \widehat{y}_T) - \left(\sum_{t=0}^{T} \alpha_t\right)^{-1} \sum_{t=0}^{T} \alpha_t f(x_t, y_t) \right|$$

$$\leq \left(2\sum_{t=0}^{T} \alpha_t\right)^{-1} (\max\{2M_1, 2M_2\} + 1)^2 + \max\{L_1\delta_C, \ L_2\delta_D\}$$

$$+ \left(\sum_{t=0}^{T} \alpha_t\right)^{-1} \max\left\{\sum_{t=0}^{T} b_{t,1}, \sum_{t=0}^{T} b_{t,2}\right\}$$

and for each $z \in C$ and each $v \in D$,

$$f(z, \widehat{y}_T) \geq f(\widehat{x}_T, \widehat{y}_T)$$

$$-\left(\sum_{t=0}^{T}\alpha_t\right)^{-1}(\max\{2M_1, 2M_2\}+1)^2 - \max\{L_1\delta_C,\ L_2\delta_D\}$$

$$-2\left(\sum_{t=0}^{T}\alpha_t\right)^{-1}\max\left\{\sum_{t=0}^{T}b_{t,1},\ \sum_{t=0}^{T}b_{t,2}\right\}$$

$$f(\widehat{x}_T, v) \le f(\widehat{x}_T, \widehat{y}_T)$$

$$+\left(\sum_{t=0}^{T}\alpha_t\right)^{-1}(\max\{2M_1, 2M_2\}+1)^2 + \max\{L_1\delta_C,\ L_2\delta_D\}$$

$$+2\left(\sum_{t=0}^{T}\alpha_t\right)^{-1}\max\left\{\sum_{t=0}^{T}b_{t,1},\ \sum_{t=0}^{T}b_{t,2}\right\}.$$

Proof By (4.11) and (4.39)–(4.41), for all integers $t \ge 0$,

$$\|x_t\| \le M_1 + 1, \quad \|y_t\| \le M_2 + 1,$$

$$B_X(x_t, \delta_C) \cap C \ne \emptyset, \quad B_Y(y_t, \delta_D) \cap D \ne \emptyset.$$

In view (4.34)–(4.36) and (4.39),

$$B_X(x_0, r_1) \subset U, \quad B_Y(y_0, r_1) \subset V.$$

It follows from (4.34)–(4.36), (4.38), (4.40), and (4.41) that

$$B_X(x_t, r_1) \subset B_X(P_{t-1}(x_{t-1} - \alpha_{t-1}\xi_{t-1}), \delta_C + r_1)$$

$$\subset B_X(P_{t-1}(x_{t-1} - \alpha_{t-1}\xi_{t-1}), 2r_1) \subset U,$$

$$B_Y(y_t, r_2) \subset B_Y(Q_{t-1}(y_{t-1} + \alpha_{t-1}\eta_{t-1}), \delta_D + r_2)$$

$$\subset B_Y(Q_{t-1}(y_{t-1} + \alpha_{t-1}\eta_{t-1}), 2r_2) \subset V.$$

Let $t \ge 0$ be an integer. Applying Lemma 3.5 with

$$r = r_1, \quad \Delta = \Delta_1, \quad P = P_t, \quad \alpha = a_t, \delta_f = \delta_{f,1},$$

$$x = x_t, \quad f = f(\cdot, y_t), \quad \xi = \xi_t, \quad y = x_{t+1},$$

we obtain that for each $z \in C$,

$$2\alpha_t(f(x_t, y_t) - f(z, y_t)) \le \|z - x_t\|^2 - \|z - x_{t+1}\|^2 + 2\alpha_t \Delta_1$$
$$+\delta_C^2 + \alpha_t^2(L_1 + \Delta_1 r^{-1}) + 2\delta_C(2M_1 + L_1 + 2 + \Delta_1 r^{-1})$$
$$+2\alpha_t \delta_{f,1}(2M_1 + L_1 + 2 + \Delta_1 r^{-1})$$
$$\le \|z - x_t\|^2 - \|z - x_{t+1}\|^2 + 2b_{t,1}.$$

Applying Lemma 3.5 with

$$r = r_2, \ \Delta = \Delta_2, \ P = Q_t, \ \alpha = a_t, \ \delta_f = \delta_{f,2},$$

$$x = y_t, \ f = -f(x_t, \cdot), \ \xi = -\eta_t, \ y = y_{t+1}$$

we obtain that for each $v \in D$,

$$2\alpha_t(f(x_t, v) - f(x_t, y_t)) \le \|v - y_t\|^2 - \|v - y_{t+1}\|^2 + 2\alpha_t \Delta_2$$
$$+\delta_D^2 + 2\delta_D(2M_2 + L_2 + 2 + \Delta r_2^{-1})$$
$$+2\alpha_t \delta_{f,2}(2M_2 + L_2 + 2 + \Delta r_2^{-1}) + \alpha_t^2(L_2 + \Delta r_2^{-1})$$
$$\le \|v - y_t\|^2 - \|v - y_{t+1}\|^2 + 2b_{t,2}.$$

Define

$$\phi(s) = 2^{-1}s^2, \ s \in R^1.$$

It is easy to see that all the assumptions of Proposition 4.1 hold and it implies Theorem 4.4.

Theorems 4.3 and 4.4 are new.

We are interested in the optimal choice of $\alpha_t, t = 0, 1, \ldots, T$. Let T be a natural number and $A_T = \sum_{t=0}^{T} \alpha_t$ be given. In order to make the best choice of $\alpha_t, t = 0, \ldots, T$, we need to minimize the function $\sum_{t=0}^{T} \alpha_t^2$ on the set

$$\left\{ \alpha = (\alpha_0, \ldots, \alpha_T) \in R^{T+1} : \alpha_i \ge 0, \ i = 0, \ldots, T, \ \sum_{i=0}^{T} \alpha_i = A_T \right\}.$$

By Lemma 2.3 of [75], this function has a unique minimizer

$$\alpha_i = (T + 1)^{-1} A_T, \ i = 0, \ldots, T.$$

Let T be a natural number and $\alpha_t = \alpha$ for all $t = 0, \ldots, T$. Now we will find the best $\alpha > 0$. In order to meet this goal we need to choose a which is a minimizer of the function

$$((T+1)\alpha)^{-1}(\max\{2M_1, 2M_2\}+1)^2$$

$$+2\alpha^{-1}(T+1)^{-1}\max\left\{\sum_{t=0}^{T}b_{t,1}, \sum_{t=0}^{T}b_{t,2}\right\}$$

$$= ((T+1)\alpha)^{-1}(\max\{2M_1, 2M_2\}+1)^2$$

$$+2\alpha^{-1}(T+1)^{-1}\max\{(T+1)(\alpha\Delta_1 + \delta_C(2M_1+3+L_1+\Delta_1 r_1^{-1}))$$

$$+2^{-1}\alpha^2(L_1+\Delta_1 r_1^{-1}) + \alpha\delta_{f,1}(2M_1+2+L_1+\Delta_1 r_1^{-1}),$$

$$(T+1)(\alpha\Delta_2 + \delta_D(2M_2+3+L_2+\Delta_2 r_2^{-1}))$$

$$+2^{-1}\alpha^2(L_2+\Delta_2 r_2^{-1}) + \alpha\delta_{f,2}(2M_2+2+L_2+\Delta_2 r_2^{-1})\}$$

$$= ((T+1)\alpha)^{-1}(\max\{2M_1, 2M_2\}+1)^2$$

$$+2\max\{\Delta_1 + \alpha^{-1}\delta_C(2M_1+3+L_1+\Delta_1 r_1^{-1})$$

$$+2^{-1}\alpha(L_1+\Delta_1 r_1^{-1}) + \delta_{f,1}(2M_1+2+L_1+\Delta_1 r_1^{-1}),$$

$$\Delta_2 + \alpha^{-1}\delta_D(2M_2+3+L_2+\Delta_2 r_2^{-1})$$

$$+2^{-1}\alpha(L_2+\Delta_2 r_2^{-1}) + \delta_{f,2}(2M_2+2+L_2+\Delta_2 r_2^{-1})\}$$

$$\leq ((T+1)\alpha)^{-1}(\max\{2M_1, 2M_2\}+1)^2$$

$$+2\max\{\Delta_1, \Delta_2\}$$

$$+2\max\{\delta_{f,1}(2M_1+2+L_1+\Delta_1 r_1^{-1}), \delta_{f,2}(2M_2+2+L_2+\Delta_2 r_2^{-1})\}$$

$$+2\alpha^{-1}\max\{\delta_C(2M_1+3+L_1+\Delta_1 r_1^{-1}), \delta_D(2M_2+3+L_2+\Delta_2 r_2^{-1})\}$$

$$+\alpha\max\{L_1+\Delta_1 r_1^{-1}, L_2+\Delta_2 r_2^{-1}\}.$$

Since T can be arbitrarily large, we need to find a minimizer of the function

$$\phi(\alpha) := 2\alpha^{-1}\max\{\delta_C(2M_1+3+L_1+\Delta_1 r_1^{-1}), \delta_D(2M_2+3+L_2+\Delta_2 r_2^{-1})\}$$

$$+\alpha\max\{L_1+\Delta_1 r_1^{-1}, L_2+\Delta_2 r_2^{-1}\}, \ \alpha > 0.$$

This function has a minimizer

$$\alpha_* = 2^{1/2}\max\{\delta_C(2M_1+3+L_1+\Delta_1 r_1^{-1}), \delta_D(2M_2+3+L_2+\Delta_2 r_2^{-1})\}^{1/2}$$

$$\times \max\{L_1+\Delta_1 r_1^{-1}, L_2+\Delta_2 r_2^{-1}\}^{-1/2}$$

and

$$\psi(\alpha_*) = 2^{3/2} \max\{\delta_C(2M_1 + 3 + L_1 + \Delta_1 r_1^{-1}), \delta_D(2M_2 + 3 + L_2 + \Delta_2 r_2^{-1})\}^{1/2}$$
$$\times \max\{L_1 + \Delta_1 r_1^{-1}, \ L_2 + \Delta_2 r_2^{-1}\}^{1/2}.$$

For the appropriate choice of T, it should be at the same order as $\max\{\delta_C, \ \delta_D\}^{-1}$.

Chapter 5
Quasiconvex Optimization

In this chapter we study an extension of the projected subgradient method for minimization of quasiconvex and nonsmooth functions, under the presence of computational errors. The problem is described by an objective function and a set of feasible points. We extend some of the results of Chapter 2 and show that our algorithm generates a good approximate solution, if all the computational errors are bounded from above by a small positive constant. Moreover, if we know computational errors for the two steps of our algorithm, we find out what an approximate solution can be obtained and how many iterates one needs for this.

5.1 Preliminaries

Let $(X, \langle \cdot, \cdot \rangle)$ be a Hilbert space with an inner product $\langle \cdot, \cdot \rangle$ which induces a complete norm $\| \cdot \|$. We use the notation and definitions introduced in Chapter 2.

For each $x \in X$ and each nonempty set $A \subset X$, set

$$d(x, A) = \inf\{\|x - y\| : y \in A\}.$$

For each $x \in X$ and each $r > 0$, set

$$B_X(x, r) = \{y \in X : \|x - y\| \leq r\}.$$

The boundary of a set $E \subset X$ is denoted by $\mathrm{bd}(E)$ and the closure of E is denoted by $\mathrm{cl}(E)$.

Let C be a closed nonempty subset of the space X and U be an open convex subset of X such that

$$C \subset U. \tag{5.1}$$

© The Editor(s) (if applicable) and The Author(s), under exclusive license to
Springer Nature Switzerland AG 2020
A. J. Zaslavski, *The Projected Subgradient Algorithm in Convex Optimization*,
SpringerBriefs in Optimization, https://doi.org/10.1007/978-3-030-60300-7_5

Suppose that $f : U \to R^1$ is a continuous function such that

$$f(\lambda x + (1 - \lambda)y) \le \max\{f(x), \ f(y)\} \tag{5.2}$$

for all $x, y \in U$ and all $\lambda \in [0, 1]$. In other words, the function f is quasiconvex. We consider the minimization problem

$$f(x) \to \min, \ x \in C.$$

It should be mentioned that quasiconvex optimization problems are studied in [19, 38–43, 77].

We suppose that

$$\inf(f, C) := \inf\{f(x) : \ x \in C\}$$

is finite and that

$$C_{min} = \{x \in C : \ f(x) = \inf(f, C)\} \ne \emptyset. \tag{5.3}$$

Fix

$$x^* \in C_{min}. \tag{5.4}$$

Set

$$S = \{x \in X : \ \|x\| = 1\}. \tag{5.5}$$

For each $x \in U$ satisfying $f(x) > \inf(f(z) : \ z \in U\}$, define

$$\partial^* f(x) = \{l \in X : \ \langle l, y - x \rangle \le 0$$
$$\text{for every } y \in U \text{ satisfying } f(y) < f(x)\}. \tag{5.6}$$

This set is a closed convex cone and it is called the subdifferential of f at the point x. Its elements are called subgradients.

Let $\epsilon \ge 0$. For each $x \in U$ satisfying $f(x) - \epsilon > \inf(f(z) : \ z \in U\}$, define

$$\partial_\epsilon^* f(x) = \{l \in X : \ \langle l, y - x \rangle \le 0$$
$$\text{for every } y \in U \text{ satisfying } f(y) < f(x) - \epsilon\}. \tag{5.7}$$

This set is a closed convex cone and it is called the ϵ-subdifferential of f at the point x. Its elements are called ϵ-subgradients.

Proposition 5.1 *Let $x \in U$, $L > 0$, $\beta > 0$, $\epsilon \geq 0$, $M_0 > 0$,*

$$x, \; x^* \in B_X(0, M_0), \tag{5.8}$$

$$f(x^*) < f(x) - \epsilon \tag{5.9}$$

and let

$$|f(z) - f(x^*)| \leq L\|z - x^*\|^\beta \text{ for all } z \in B_X(0, 3M_0 + 1) \cap U. \tag{5.10}$$

Then for all $g \in \partial_\epsilon^* f(x) \cap S$,

$$f(x) - f(x^*) \leq L\langle g, x - x^*\rangle^\beta + \epsilon.$$

Proof Let

$$g \in \partial_\epsilon^* f(x) \cap S. \tag{5.11}$$

In view of the continuity and quasiconvexity of the function f, the set

$$E := \{z \in U : \; f(z) < f(x) - \epsilon\} \tag{5.12}$$

is open and convex. By (5.9) and (5.12),

$$x^* \in E. \tag{5.13}$$

For the continuity of f, (5.12) and (5.13) imply that $\mathrm{bd}(E) \neq \emptyset$ and that

$$\{\lambda x + (1 - \lambda)x^* : \; \lambda \in [0, 1]\} \cap \mathrm{bd}(E) \neq \emptyset. \tag{5.14}$$

Set

$$r = \inf\{\|x^* - u\| : \; u \in \mathrm{bd}(E) \cap U\}. \tag{5.15}$$

It follows from (5.8), (5.14), and (5.15) that

$$r \leq \|x^* - x\| \leq 2M_0. \tag{5.16}$$

There exists a sequence $\{u_k\}_{k=1}^\infty \subset \mathrm{bd}(E) \cap U$ such that

$$\|x^* - u_k\| \leq r + k^{-1}, \; k = 1, 2, \ldots. \tag{5.17}$$

The openness of E implies that

$$f(x) \leq f(u) + \epsilon \text{ for all } u \in \mathrm{bd}(E) \cap U. \tag{5.18}$$

In view of (5.18), for every integer $k \geq 1$,

$$f(x) - f(x^*) \leq f(u_k) - f(x^*) + \epsilon. \tag{5.19}$$

Let $k \geq 1$ be an integer. By (5.8), (5.16), and (5.17),

$$\|u_k\| \leq r + k^{-1} + \|x^*\| \leq 3M_0 + 1. \tag{5.20}$$

It follows from (5.8), (5.10), (5.17), (5.19), and (5.20) that

$$f(x) - f(x^*) \leq f(u_k) - f(x^*) + \epsilon \leq L\|u_k - x^*\|^\beta + \epsilon$$
$$\leq L(r + k^{-1})^\beta + \epsilon$$

and

$$f(x) - f(x^*) \leq Lr^\beta + \epsilon. \tag{5.21}$$

In view of (5.15),

$$\{z \in U : \|z - x^*\| < r\} \subset E. \tag{5.22}$$

Let k be a natural number. By (5.11) and (5.22),

$$x^* + (1 - k^{-1})rg \in E. \tag{5.23}$$

It follows from (5.11), (5.12), and (5.23) that

$$(1 - k^{-1})r - \langle g, x - x^* \rangle = (1 - k^{-1})r\|g\|^2 - \langle g, x - x^* \rangle$$
$$= \langle g, x^* + (1 - k^{-1})rg - x \rangle \leq 0$$

and

$$r \leq \langle g, x - x^* \rangle.$$

Together with (5.21) this implies that

$$f(x) - f(x^*) \leq L\langle g, x - x^* \rangle^\beta + \epsilon.$$

Proposition 5.1 is proved.

This chapter contains two main results, Theorems 5.3 and 5.4, which are proved in Sections 5.3 and 5.4, respectively. These results are new and their proofs are based on our main Lemma 5.2 proved in Section 5.2.

5.2 The Main Lemma

Lemma 5.2 *Let $M_0 > 0$, $L > 0$, $\beta > 0$, $\delta_f, \delta_C \in (0, 1]$, $\Delta \geq 0$, $\alpha \in (0, 1]$, $x \in U$ and let a mapping $P : X \to X$ satisfy*

$$Py = y, \ y \in C, \tag{5.24}$$

$$\|Py - z\| \leq \|y - z\| \text{ for all } y \in X \text{ and all } z \in C. \tag{5.25}$$

Assume that

$$\|x^*\| \leq M_0, \ \|x\| \leq M_0, \tag{5.26}$$

$$|f(z) - f(x^*)| \leq L\|z - x^*\|^\beta, \ z \in B_X(0, 3M_0 + 1) \cap U, \tag{5.27}$$

$$f(x) > \inf(f, C) + \Delta, \tag{5.28}$$

$\xi \in X$ satisfies

$$B_X(\xi, \delta_f) \cap \partial_\Delta^* f(x) \cap S \neq \emptyset \tag{5.29}$$

and that $y \in U$ satisfies

$$\|y - P(x - \alpha\xi)\| \leq \delta_C. \tag{5.30}$$

Then

$$\|y - x^*\|^2 \leq \|x - x^*\|^2 + \alpha^2(1 + \delta_f^2) + 2\alpha\delta_f(2M_0 + 1)$$
$$+\delta_C^2 + 2\delta_C(2M_0 + 2) - 2\alpha L^{-1/\beta}(f(x) - \inf(f, C) - \Delta)^{1/\beta}.$$

Proof In view of (5.29), there exists

$$g \in \partial_\Delta^* f(x) \cap S \tag{5.31}$$

such that

$$\|g - \xi\| \leq \delta_f. \tag{5.32}$$

Proposition 5.1 and (5.31) imply that

$$f(x) - \inf(f, C) \leq L\langle g, x - x^*\rangle^\beta + \Delta. \tag{5.33}$$

By (5.3), (5.4), (5.25), (5.26), and (5.30)–(5.32),

$$
\begin{aligned}
\|y - x^*\|^2 &\le \|y - P(x - \alpha\xi) + P(x - \alpha\xi) - x^*\|^2 \\
&\le \|y - P(x - \alpha\xi)\|^2 + \|P(x - \alpha\xi) - x^*\|^2 \\
&\quad + 2\|y - P(x - \alpha\xi)\|\|P(x - \alpha\xi) - x^*\| \\
&\le \delta_C^2 + \|P(x - \alpha\xi) - x^*\|^2 + 2\delta_C\|P(x - \alpha\xi) - x^*\| \\
&\le \delta_C^2 + \|x - \alpha\xi - x^*\|^2 + 2\delta_C\|x - \alpha\xi - x^*\| \\
&\le \delta_C^2 + \|x - \alpha\xi - x^*\|^2 + 2\delta_C(2M_0 + 2).
\end{aligned} \tag{5.34}
$$

It follows from (5.26), (5.28), and (5.31)–(5.33) that

$$
\begin{aligned}
\|x - \alpha\xi - x^*\|^2 &= \|x - \alpha g + (\alpha g - \alpha\xi) - x^*\|^2 \\
&\le \|x - \alpha\xi - x^*\|^2 + \alpha^2\|g - \xi\|^2 + 2\alpha\|g - \xi\|\|x - \alpha\xi - x^*\| \\
&\le \|x - \alpha g - x^*\|^2 + \alpha^2\delta_f^2 + 2\alpha\delta_f(2M_0 + 1) \\
&\le \alpha^2\delta_f^2 + 2\alpha\delta_f(2M_0 + 1) + \|x - x^*\|^2 + \alpha^2 - 2\alpha\langle g, x - x^*\rangle \\
&\le \alpha^2\delta_f^2 + 2\alpha\delta_f(2M_0+1) + \|x-x^*\|^2 + \alpha^2 - 2\alpha L^{-1/\beta}(f(x) - \inf(f, C) - \Delta)^{1/\beta}).
\end{aligned}
$$

Together with (5.34) this implies that

$$
\begin{aligned}
\|y - x^*\|^2 &\le \|x - x^*\|^2 + \alpha^2(1 + \delta_f^2) + 2\alpha\delta_f(2M_0 + 1) \\
&\quad - 2\alpha L^{-1/\beta}(f(x) - \inf(f, C) - \Delta)^{1/\beta}) + \delta_C^2 + 2\delta_C(2M_0 + 2).
\end{aligned}
$$

Lemma 5.2 is proved.

5.3 Optimization on Bounded Sets

Denote by $\mathcal{M}$ the set of all mappings $P : X \to C$ such that

$$
Pz = z, \quad z \in C, \tag{5.35}
$$

$$
\|Px - z\| \le \|x - z\| \text{ for all } x \in X \text{ and all } z \in C. \tag{5.36}
$$

Theorem 5.3 *Let $M > 1$, $L > 0$, $\beta > 0$, $\delta_f, \delta_C \in (0, 1]$, $\Delta \ge 0$, T be a natural number, $\{P_t\}_{t=0}^{T-1} \subset \mathcal{M}$, $\{\alpha_t\}_{t=0}^{T-1} \subset (0, 1]$,*

$$C \subset B_X(0, M - 1), \tag{5.37}$$

$$|f(z) - f(x^*)| \leq L\|z - x^*\|^{\beta}, \ z \in B_X(0, 3M + 1) \cap U \tag{5.38}$$

and let

$$\epsilon = \Delta + \left(2^{-1}L^{1/\beta}(2M + 3)^2 \left(\sum_{t=0}^{T-1} \alpha_t\right)^{-1}\right.$$

$$\left. + L^{1/\beta}\left(\sum_{t=0}^{T-1} \alpha_t\right)^2 \left(\sum_{t=0}^{T-1} \alpha_t\right)^{-1} + L^{1/\beta}\delta_f(2M + 1) + L^{1/\beta}\delta_C(2M + 3)\left(\sum_{t=0}^{T-1} \alpha_t\right)^{-1}\right)^{\beta}.$$

$$\tag{5.39}$$

Assume that $\{x_t\}_{t=0}^T \subset U$, $\{\xi_t\}_{t=0}^{T-1} \subset X$,

$$B(x_0, \delta_C) \cap C \neq \emptyset \tag{5.40}$$

and that for all $t = 0, \ldots, T - 1$,

$$B_X(\xi_t, \delta_f) \cap \partial_{\Delta}^* f(x_t) \cap S \neq \emptyset, \tag{5.41}$$

$$\|x_{t+1} - P_t(x_t - \alpha_t\xi_t)\| \leq \delta_C. \tag{5.42}$$

Then

$$\min\{f(x_t) : t = 0, \ldots, T - 1\} \leq \inf(f, C) + \epsilon. \tag{5.43}$$

Proof Clearly, $\partial f_{\Delta}^*(z)$ is well-defined if

$$f(z) - \Delta > \inf\{f(x) : x \in U\}.$$

We may assume without loss of generality that

$$f(x_t) - \Delta > \inf\{f(x) : x \in U\}, \ t = 0, \ldots, T - 1.$$

We show that (5.43) holds. Assume the contrary. Then

$$f(x_t) > \epsilon + \inf(f, C) = \epsilon + f(x^*), \ t = 0, \ldots, T - 1. \tag{5.44}$$

Let $t \in \{0, \ldots, T - 1\}$. By (5.35)–(5.42), (5.44), and Lemma 5.2 applied with $\alpha = \alpha_t$, $P = P_t$, $x = x_t$, $\xi = \xi_t$, $y = x_{t+1}$, and $M_0 = M$,

$$\|x_{t+1} - x^*\|^2 \leq \|x_t - x^*\|^2 + 2\alpha_t^2 + 2\alpha_t\delta_f(2M + 1)$$

$$-2\alpha_t L^{-1/\beta}(\epsilon - \Delta)^{1/\beta} + 2\delta_C(2M + 3)$$

and

$$\|x_t - x^*\|^2 - \|x_{t+1} - x^*\|^2$$
$$\geq 2\alpha_t L^{-1/\beta}(\epsilon - \Delta)^{1/\beta}$$
$$-2\alpha_t^2 - 2\alpha_t \delta_f(2M + 1) - 2\delta_C(2M + 3). \tag{5.45}$$

It follows from (5.37), (5.40), and (5.45) that

$$(2M + 2)^2 > \|x_0 - x^*\|^2 \geq \|x_0 - x^*\|^2 - \|x_T - x^*\|^2$$
$$= \sum_{t=0}^{T-1}(\|x_t - x^*\|^2 - \|x_{t+1} - x^*\|^2)$$
$$\geq 2L^{-1/\beta}(\epsilon - \Delta)^{1/\beta}\sum_{t=0}^{T-1}\alpha_t - 2\sum_{t=0}^{T-1}\alpha_t^2$$
$$-2\delta_f(2M + 1)\sum_{t=0}^{T-1}\alpha_t - 2\delta_C(2M + 3)T$$

and

$$(\epsilon - \Delta)^{1/\beta} < 2^{-1}L^{1/\beta}(2M + 3)^2\left(\sum_{t=0}^{T-1}\alpha_t\right)^{-1} + L^{1/\beta}\left(\sum_{t=0}^{T-1}\alpha_t^2\right)\left(\sum_{t=0}^{T-1}\alpha_t\right)^{-1}$$
$$+ L^{1/\beta}\delta_f(2M + 1) + L^{1/\beta}\delta_C(2M + 3)T\left(\sum_{t=0}^{T-1}\alpha_t\right)^{-1}.$$

This contradicts (5.39). The contradiction we have reached completes the proof of Theorem 5.3.

Let T be a natural number and $\sum_{i=0}^{T-1}\alpha_t = A$ is Theorem 5.3 are given. We need to choose $\alpha_t \geq 0$, $t = 0, \ldots, T - 1$ in order to minimize ϵ. Clearly, we need to minimize the function $\sum_{i=0}^{T-1}\alpha_t^2$ over the set

$$\left\{(\alpha_0, \ldots, \alpha_{T-1}) \in R^T : \alpha_i \geq 0, \ i = 0, \ldots, T - 1, \ \sum_{i=0}^{T-1}\alpha_t = A\right\}.$$

By Lemma 2.3 of [75], we should choose $\alpha_t = AT^{-1}, t = 0, \ldots, T - 1$.
Assume that $\alpha_t = \alpha > 0, \ t = 0, \ldots, T - 1$. In this case in view of (5.39),

$$\epsilon = \Delta + (2^{-1}L^{1/\beta}(2M + 3)^2(\alpha T)^{-1} + L^{1/\beta}\alpha + L^{1/\beta}\delta_f(2M + 1)$$
$$+L^{1/\beta}\delta_C(2M + 3)\alpha^{-1})^\beta.$$

We need to choose $\alpha > 0$ in order to minimize ϵ. Since a natural number T can be arbitrarily large, we need to minimize the function

$$\alpha + \delta_C(2M + 3)\alpha^{-1}, \ \alpha > 0.$$

Clearly, its minimizer is

$$\alpha = (\delta_C(2M + 3))^{1/2}.$$

In this case

$$\epsilon = \Delta + (2^{-1}L^{1/\beta}(2M + 3)^2(\delta_C(2M + 3))^{-1/2}T^{-1}$$
$$+L^{1/\beta}(\delta_C(2M + 3)) + L^{1/\beta}\delta_f(2M + 1) + L^{1/\beta}(\delta_C(2M + 3))^{1/2})^\beta.$$

Evidently, T should be at the same order as $\lfloor \delta_C^{-1} \rfloor$.

5.4 Optimization on Unbounded Sets

We continue to use the definitions and notation introduced in Section 5.1. Recall that $\mathcal{M}$ is the set of all mappings $P : X \to C$ which satisfy (5.35) and (5.36).
Assume that

$$\lim_{x\in U, \ \|x\|\to\infty} f(x) = \infty. \tag{5.46}$$

It means that for each $M_0 > 0$, there exists $M_1 > 0$ such that if a point $x \in U$ satisfies $\|x\| \geq M_1$, then $f(x) > M_0$.
Fix

$$\theta_0 \in C. \tag{5.47}$$

Set

$$U_0 = \{x \in U : \ f(x) \le f(\theta_0) + 4\}. \tag{5.48}$$

By (5.46) and (5.48), there exists $\bar{M} > 1$ such that

$$U_0 \subset B_X(0, \bar{M}). \tag{5.49}$$

Assume that there exists a number $\bar{L} > 1$ such that

$$|f(z) - f(x^*)| \le \bar{L}\|z - x^*\|^\beta, \ z \in B_X(0, \bar{M} + 4) \cap U. \tag{5.50}$$

Theorem 5.4 *Let* $M \ge \bar{M} + 4$, $L \ge \bar{L} + 1$, $\beta > 0$, $\delta_f, \delta_C \in (0, 1]$, $\Delta \in (0, 1]$,

$$0 < \alpha \le 4^{-1}L^{-1/\beta}, \ \delta_C \le (6M + 3)^{-1}4^{-1}L^{-1/\beta}\alpha, \ \delta_f \le (6M + 3)^{-1}4^{-1}L^{-1/\beta} \tag{5.51}$$

and

$$|f(z) - f(x^*)| \le L\|z - x^*\|^\beta, \ z \in B_X(0, 9M + 4) \cap U. \tag{5.52}$$

Assume that T *is a natural number,*

$$\{P_t\}_{t=0}^{T-1} \subset \mathcal{M}, \tag{5.53}$$

$\{x_t\}_{t=0}^{T} \subset U, \{\xi_t\}_{t=0}^{T-1} \subset X,$

$$\|x_0\| \le M, \tag{5.54}$$

$$B(x_0, \delta_C) \cap C \ne \emptyset \tag{5.55}$$

and that for all $t = 0, \ldots, T - 1$,

$$B_X(\xi_t, \delta_f) \cap \partial_\Delta^* f(x_t) \cap S \ne \emptyset, \tag{5.56}$$

$$\|x_{t+1} - P_t(x_t - \alpha\xi_t)\| \le \delta_C. \tag{5.57}$$

Then

$$\min\{f(x_t) : \ t = 0, \ldots, T - 1\} \le \inf(f, C) + \Delta$$
$$+ L(2M^2(\alpha T)^{-1} + \alpha + \delta_C\alpha^{-1}(6M + 2) + \delta_f(6M + 2))^\beta. \tag{5.58}$$

Proof Clearly, $\partial f_\Delta^*(z)$ is well-defined if

$$f(z) - \Delta > \inf\{f(x) : \ x \in U\}.$$

We may assume without loss of generality that

$$f(x_t) - \Delta > \inf\{f(x) : x \in U\}, \ t = 0, \ldots, T - 1.$$

We show that (5.58) holds. Assume the contrary. Then for all $t = 0, \ldots, T - 1$,

$$f(x_t) - f(x^*) > \Delta + L(2M^2(\alpha T)^{-1} + \alpha + \delta_C \alpha^{-1}(6M + 2) + \delta_f (6M + 2))^\beta.$$
(5.59)

In view of (5.3), (5.4), (5.49), and (5.54),

$$\|x_0 - x^*\| \le M + \bar{M}.$$

By induction we show that for all $t = 0, \ldots, T$,

$$\|x_t - x^*\| \le M + \bar{M}.$$
(5.60)

Clearly, (5.60) is true for $t = 0$. Assume that $t \in \{0, \ldots, T - 1\}$ and (5.60) is true. There are two cases:

$$f(x_t) \le \inf(f, C) + 4;$$
(5.61)

$$f(x_t) > \inf(f, C) + 4.$$
(5.62)

Assume that (5.61) holds. By (5.47)–(5.49) and (5.61),

$$\|x_t\| \le \bar{M}.$$
(5.63)

It follows from (5.3), (5.4), (5.35), (5.36), (5.47)–(5.49), (5.56), (5.57), and (5.63) that

$$\|x_{t+1} - x^*\| \le \|x_{t+1} - P_t(x_t - \alpha \xi_t)\| + \|P_t(x_t - \alpha \xi_t) - x^*\|$$
$$\le \delta_C + \|x_t - \alpha \xi_t - x^*\| \le 2\bar{M} + 3 \le M + \bar{M}.$$

Assume that (5.62) holds. By (5.60) which implies $\|x_t\| \le 3M$, (5.51)–(5.53), (5.56), (5.57), (5.62), and Lemma 5.2 applied with $P = P_t$, $x = x_t$, $\xi = \xi_t$, $y = x_{t+1}$, and $M_0 = 3M$,

$$\|x_{t+1} - x^*\|^2 \le \|x_t - x^*\|^2 + 2\alpha^2 + 2\alpha \delta_f (6M + 1)$$
$$-2\alpha L^{-1/\beta} 3^{1/\beta} + 2\delta_C (6M + 3)$$

and

$$\|x_{t+1} - x^*\| \le \|x_t - x^*\| \le M + \bar{M}.$$

Thus in both cases

$$\|x_{t+1} - x^*\| \le M + \bar{M}.$$

Therefore (5.60) holds for all $t = 0, \ldots, T$. In view of (5.3), (5.4), (5.47)–(5.49), and (5.60),

$$\|x_t\| \le 3M, \; t = 0, \ldots, T. \tag{5.64}$$

Let $t \in \{0, \ldots, T - 1\}$. By (5.35), (5.36), (5.52), (5.56), (5.57), (5.59), (5.64), and Lemma 5.2 applied with $P = P_t$, $x = x_t$, $\xi = \xi_t$, $y = x_{t+1}$, and $M_0 = 3M$,

$$\|x_{t+1} - x^*\|^2 \le \|x_t - x^*\|^2 + 2\alpha^2 + 2\alpha\delta_f(6M + 1)$$
$$+\delta_C^2 + 2\delta_C(6M + 2) - 2\alpha L^{-1/\beta}(f(x_t) - f(x^*) - \Delta)^{1/\beta}$$
$$\le \|x_t - x^*\|^2 + 2\alpha^2 + 2\alpha\delta_f(6M + 1)$$
$$+2\delta_C(6M + 3) - 2\alpha L^{-1/\beta}(f(x_t) - f(x^*) - \Delta)^{1/\beta}. \tag{5.65}$$

It follows from (5.60) and (5.65) that

$$(2M)^2 \ge \|x_0 - x^*\|^2 \ge \|x_0 - x^*\|^2 - \|x_T - x^*\|^2$$
$$= \sum_{t=0}^{T-1} (\|x_t - x^*\|^2 - \|x_{t+1} - x^*\|^2)$$
$$\ge 2\alpha \sum_{t=0}^{T-1} (L^{-1/\beta}(f(x_t) - f(x^*) - \Delta)^{1/\beta}$$
$$-\alpha - \alpha^{-1}\delta_C(6M + 2) - \delta_f(6M + 1))$$
$$\ge 2\alpha T((\min\{f(x_t) : t = 0, \ldots, T - 1\} - f(x^*) - \Delta)^{1/\beta} L^{-1/\beta}$$
$$-\alpha - \alpha^{-1}\delta_C(6M + 2) - \delta_f(6M + 1))$$

and

$$(\min\{f(x_t) : t = 0, \ldots, T - 1\} - f(x^*) - \Delta)^{1/\beta}$$
$$< L^{1/\beta}(2M^2(T\alpha)^{-1} + \alpha + \alpha^{-1}\delta_C(6M + 2) + \delta_f(6M + 1)).$$

This implies that

$$\min\{f(x_t) : t = 0, \ldots, T - 1\} < \inf(f, C) + \Delta$$
$$+L(2M^2(T\alpha)^{-1} + \alpha + \alpha^{-1}\delta_C(6M + 2) + \delta_f(6M + 1))^\beta.$$

This contradicts (5.59). The contradiction we have reached proves Theorem 5.4.

In order to minimize the right-hand side of (5.58), we should choose

$$\alpha = (\delta_C(6M + 2))^{1/2}.$$

In this case T should be at the same order as $\lfloor \delta_C^{-1} \rfloor$. Of course, we need that (5.51) is true. This leads us to the following condition on δ_C:

$$\delta_C \leq (6M + 3)^{-1} 4^{-2} L^{-2\beta}.$$

Together with our choice of α and the condition above on δ_C, Theorem 5.4 implies the following result.

Theorem 5.5 *Let* $M \geq \bar{M} + 4$, $L \geq \bar{L} + 1$, $\beta > 0$, $\delta_f, \delta_C \in (0, 1]$, $\Delta \in (0, 1]$,

$$|f(z) - f(x^*)| \leq L\|z - x^*\|^\beta, \quad z \in B_X(0, 9M + 4) \cap U,$$

$$\delta_C \leq (6M + 3)^{-1} 4^{-2} L^{-2/\beta}, \quad \delta_f \leq (6M + 3)^{-1} 4^{-1} L^{-1/\beta}$$

and

$$\alpha = (\delta_C(6M + 2))^{1/2}.$$

Assume that T *is a natural number,*

$$\{P_t\}_{t=0}^{T-1} \subset \mathcal{M},$$

$$\{x_t\}_{t=0}^T \subset U, \{\xi_t\}_{t=0}^{T-1} \subset X,$$

$$\|x_0\| \leq M,$$

$$B(x_0, \delta_C) \cap C \neq \emptyset$$

and that for all $t = 0, \ldots, T - 1$,

$$B_X(\xi_t, \delta_f) \cap \partial_\Delta^* f(x_t) \cap S \neq \emptyset,$$

$$\|x_{t+1} - P_t(x_t - \alpha_t \xi_t)\| \leq \delta_C.$$

Then

$$\min\{f(x_t) : t = 0, \ldots, T - 1\} \leq \inf(f, C) + \Delta$$
$$+ L(2M^2(T(\delta_C(6M + 2))^{1/2})^{-1}) + (\delta_C(6M + 2))^{1/2}$$
$$+ (\delta_C(6M + 3))^{1/2} + \delta_f(6M + 2))^\beta.$$

References

1. Alber YI (1971) On minimization of smooth functional by gradient methods. USSR Comp Math Math Phys 11:752–758
2. Alber YI, Iusem AN, Solodov MV (1997) Minimization of nonsmooth convex functionals in Banach spaces. J Convex Anal 4:235–255
3. Alber YI, Iusem AN, Solodov MV (1998) On the projected subgradient method for nonsmooth convex optimization in a Hilbert space. Math Program 81:23–35
4. Aleyner A, Reich S (2008) Block-iterative algorithms for solving convex feasibility problems in Hilbert and Banach spaces. J Math Anal Appl 343:427–435
5. Al-Mazrooei AE, Latif A, Qin X, Yao J-C (2019) Fixed point algorithms for split feasibility problems. Fixed Point Theory 20:245–254
6. Alsulami SM, Takahashi W (2015) Iterative methods for the split feasibility problem in Banach spaces. J Nonlinear Convex Anal 16:585–596
7. Barty K, Roy J-S, Strugarek C (2007) Hilbert-valued perturbed subgradient algorithms. Math Oper Res 32:551–562
8. Bauschke HH, Borwein JM (1996) On projection algorithms for solving convex feasibility problems. SIAM Rev 38:367–426
9. Bauschke HH, Koch, VR (2015) Projection methods: Swiss army knives for solving feasibility and best approximation problems with half-spaces. Contemp Math 636:1–40
10. Bauschke H, Wang C, Wang X, Xu J (2015) On subgradient projectors. SIAM J Optim 25:1064–1082
11. Beck A, Teboulle M (2003) Mirror descent and nonlinear projected subgradient methods for convex optimization. Oper Res Lett 31:167–175
12. Burachik RS, Grana Drummond LM, Iusem AN, Svaiter BF (1995) Full convergence of the steepest descent method with inexact line searches. Optimization 32:137–146
13. Butnariu D, Davidi, R, Herman GT, Kazantsev IG (2007) Stable convergence behavior under summable perturbations of a class of projection methods for convex feasibility and optimization problems. IEEE J Select Top Signal Process 1:540–547
14. Butnariu D, Reich S, Zaslavski AJ (2006) Convergence to fixed points of inexact orbits of Bregman-monotone and of nonexpansive operators in Banach spaces. In: Fixed point theory and its applications. Yokohama Publishers, Mexico, pp 11–32
15. Butnariu D, Reich S, Zaslavski AJ (2008) Stable convergence theorems for infinite products and powers of nonexpansive mappings. Numer Func Anal Optim 29:304–323

© The Editor(s) (if applicable) and The Author(s), under exclusive license to
Springer Nature Switzerland AG 2020
A. J. Zaslavski, *The Projected Subgradient Algorithm in Convex Optimization*,
SpringerBriefs in Optimization, https://doi.org/10.1007/978-3-030-60300-7

16. Ceng LC, Hadjisavvas N, Wong NC (2010) Strong convergence theorem by a hybrid extragradient-like approximation method for variational inequalities and fixed point problems. J Glob Optim 46:635–646
17. Ceng LC, Wong NC, Yao JC (2015) Hybrid extragradient methods for fiinding minimum norm solutions of split feasibility problems. J Nonlinear Convex Anal 16:1965–1983
18. Censor Y, Cegielski A (2015) Projection methods: an annotated bibliography of books and reviews. Optimization 64:2343–2358
19. Censor Y, Segal A (2006) Algorithms for the quasiconvex feasibility problem. J Comput Appl Math 185:34–50
20. Censor Y, Segal A (2010) On string-averaging for sparse problems and on the split common fixed point problem. Contemp Math 513:125–142
21. Censor Y, Tom E (2003) Convergence of string-averaging projection schemes for inconsistent convex feasibility problems. Optim Methods Softw 18:543–554
22. Censor Y, Zaslavski AJ (2013) Convergence and perturbation resilience of dynamic string-averaging projection methods. Comput Optim Appl 54:65–76
23. Censor Y, Zaslavski AJ (2014) String-averaging projected subgradient methods for constrained minimization. Optim Methods Softw 29:658–670
24. Censor Y, Elfving T, Herman GT (2001) Averaging strings of sequential iterations for convex feasibility problems. In: Butnariu D, Censor Y, Reich S (eds). Inherently parallel algorithms in feasibility and optimization and their applications. North-Holland, Amsterdam, pp 101–113
25. Censor Y, Davidi R, Herman GT (2010) Perturbation resilience and superiorization of iterative algorithms. Inverse Probl 26:12 pp
26. Censor Y, Gibali A, Reich S (2011) The subgradient extragradient method for solving variational inequalities in Hilbert space. J Optim Theory Appl 148:318–335
27. Censor Y, Chen W, Combettes PL, Davidi R, Herman GT (2012) On the effectiveness of projection methods for convex feasibility problems with linear inequality constraints. Comput Optim Appl 51:1065–1088
28. Censor Y, Gibali A, Reich S, Sabach S (2012) Common solutions to variational inequalities. Set-Valued Var Anal 20:229–247
29. Censor Y, Davidi R, Herman GT, Schulte RW, Tetruashvili L (2014) Projected subgradient minimization versus superiorization. J Optim Theory Appl 160:730–747
30. Chadli O, Konnov IV, Yao JC (2004) Descent methods for equilibrium problems in a Banach space. Comput Math Appl 48:609–616
31. Combettes PL (1996) The convex feasibility problem in image recovery. Adv Imag Electron Phys 95:155–270
32. Combettes PL (1997) Hilbertian convex feasibility problems: convergence of projection methods. Appl Math Optim 35:311–330
33. Demyanov VF, Vasilyev LV (1985) Nondifferentiable optimization. Optimization Software, New York
34. Gibali A, Jadamba B, Khan AA, Raciti F, Winkler B (2016) Gradient and extragradient methods for the elasticity imaging inverse problem using an equation error formulation: a comparative numerical study. Nonlinear Anal Optim Contemp Math 659:65–89
35. Griva I (2018) Convergence analysis of augmented Lagrangian-fast projected gradient method for convex quadratic problems. Pure Appl Funct Anal 3:417–428
36. He H, Xu H-K (2017) Splitting methods for split feasibility problems with application to Dantzig selectors. Inverse Probl 33:28 pp
37. Hiriart-Urruty J-B, Lemarechal C (1993) Convex analysis and minimization algorithms. Springer, Berlin
38. Hishinuma K, Iiduka H (2020) Fixed point quasiconvex subgradient method. Eur J Oper Res 282:428–437
39. Hu Y, Yang X, Sim C-K (2015) Inexact subgradient methods for quasi-convex optimization problems. Eur J Oper Res 240:315–327
40. Hu Y, Yu CKW, Li C (2016) Stochastic subgradient method for quasi-convex optimization problems. J Nonlinear Convex Anal 17:711–724

41. Hu Y, Yu CKW, Li C, Yang X (2016) Conditional subgradient methods for constrained quasi-convex optimization problems. J Nonlinear Convex Anal 17:2143–2158
42. Kiwiel KC (2001) Convergence and efficiency of subgradient methods for quasiconvex minimization. Math Programm 90:1–25
43. Konnov IV (2003) On convergence properties of a subgradient method. Optim Methods Softw 18:53–62
44. Konnov IV (2009) A descent method with inexact linear search for mixed variational inequalities. Russ Math (Iz VUZ) 53:29–35
45. Konnov IV (2018) Simplified versions of the conditional gradient method. Optimization 67:2275–2290
46. Korpelevich GM (1976) The extragradient method for finding saddle points and other problems. Ekon Matem Metody 12:747–756
47. Liu L, Qin X, Yao J-C (2019) A hybrid descent method for solving a convex constrained optimization problem with applications. Math Methods Appl Sci 42:7367–7380
48. Mordukhovich BS (2006) Variational analysis and generalized differentiation I: Basic theory. Springer, Berlin
49. Mordukhovich BS, Nam NM (2014) An easy path to convex analysis and applications. Morgan & Clayton Publishes, San Rafael
50. Nadezhkina N, Wataru T (2004) Modified extragradient method for solving variational inequalities in real Hilbert spaces. Nonlinear analysis and convex analysis. Yokohama Publishers, Yokohama, pp 359–366
51. Nedic A, Ozdaglar A (2009) Subgradient methods for saddle-point problems. J Optim Theory Appl 142:205–228
52. O'Hara JG, Pillay P, Xu HK (2006) Iterative approaches to convex feasibility problems in Banach spaces. Nonlinear Anal 64:2022–2042
53. Pallaschke D, Recht P (1985) On the steepest–descent method for a class of quasidifferentiable optimization problems. In: Nondifferentiable optimization: motivations and applications (Sopron, 1984). Lecture notes in economics and mathematical systems, vol 255. Springer, Berlin, pp 252–263
54. Polyak BT (1987) Introduction to optimization. Optimization Software, New York
55. Polyak RA (2015) Projected gradient method for non-negative least squares. Contemp. Math. 636:167–179
56. Qin X, Cho SY, Kang SM (2011) An extragradient-type method for generalized equilibrium problems involving strictly pseudocontractive mappings. J Global Optim 49:679–693
57. Reich S, Zaslavski AJ (2014) Genericity in nonlinear analysis. Developments in mathematics. Springer, New York
58. Shor NZ (1985) Minimization methods for non-differentiable functions. Springer, Berlin
59. Solodov MV, Zavriev SK (1998) Error stability properties of generalized gradient-type algorithms. J Optim Theory Appl 98:663–680
60. Su M, Xu H-K (2010) Remarks on the gradient-projection algorithm. J Nonlinear Anal Optim 1:35–43
61. Takahashi W (2009) Introduction to nonlinear and convex analysis. Yokohama Publishers, Yokohama
62. Takahashi W (2014) The split feasibility problem in Banach spaces. J Nonlinear Convex Anal 15:1349–1355
63. Takahashi W, Wen C-F, Yao J-C (2020) Strong convergence theorem for split common fixed point problem and hierarchical variational inequality problem in Hilbert spaces. J Nonlinear Convex Anal 21:251–273
64. Thuy LQ, Wen C-F, Yao J-C, Hai TN (2018) An extragradient-like parallel method for pseudomonotone equilibrium problems and semigroup of nonexpansive mappings. Miskolc Math Notes 19:1185–1201
65. Wang H, Xu H-K (2018) A note on the accelerated proximal gradient method for nonconvex optimization. Carpathian J Math 34:449–457

66. Xu H-K (2011) Averaged mappings and the gradient-projection algorithm. J Optim Theory Appl 150:360–378
67. Xu H-K (2017) Bounded perturbation resilience and superiorization techniques for the projected scaled gradient method. Inverse Probl 33:19 pp
68. Yao Y, Postolache M, Yao J-C (2019) Convergence of an extragradient algorithm for fixed point and variational inequality problems. J Nonlinear Convex Anal 20:2623–2631
69. Yao Y, Qin X, Yao J-C (2018) Constructive approximation of solutions to proximal split feasibility problems. J Nonlinear Convex Anal 19:2165–2175
70. Yao Y, Qin X, Yao J-C (2019) Convergence analysis of an inertial iterate for the proximal split feasibility problem. J Nonlinear Convex Anal 20:489–498
71. Zaslavski AJ (2010) The projected subgradient method for nonsmooth convex optimization in the presence of computational errors. Numer Funct Anal Optim 31:616–633
72. Zaslavski AJ (2012) The extragradient method for convex optimization in the presence of computational errors. Numer Funct Anal Optim 33:1399–1412
73. Zaslavski AJ (2012) The extragradient method for solving variational inequalities in the presence of computational errors. J Optim Theory Appl 153:602–618
74. Zaslavski AJ (2013) The extragradient method for finding a common solution of a finite family of variational inequalities and a finite family of fixed point problems in the presence of computational errors. J Math Anal Appl 400:651–663
75. Zaslavski AJ (2016) Numerical optimization with computational errors. Springer, Cham
76. Zaslavski AJ (2016) Approximate solutions of common fixed point problems. Springer optimization and its applications. Springer, Cham
77. Zaslavski AJ (2020) Convex optimization with computational errors. Springer optimization and its applications. Springer, Cham
78. Zeng LC, Yao JC (2006) Strong convergence theorem by an extragradient method for fixed point problems and variational inequality problems. Taiwanese J Math 10:1293–1303

Printed in the United States
By Bookmasters